Transportation Technologies

Transportation Technologies

Edited by:

Zoran Gacovski

ARCLER
PRESS

www.arclerpress.com

Transportation Technologies

Zoran Gacovski

Arcler Press
224 Shoreacres Road
Burlington, ON L7L 2H2
Canada
www.arclerpress.com
Email: orders@arclereducation.com

ISBN: 978-1-77407-774-0 (Hardcover)

DECLARATION

ABOUT THE EDITOR

Dr. Zoran Gacovski has earned his PhD degree at Faculty of Electrical engineering, Skopje. His research interests include Intelligent systems and Software engineering, fuzzy systems, graphical models (Petri, Neural and Bayesian networks), and IT security. He has published over 50 journal and conference papers, and he has been reviewer of renowned Journals. Currently, he is a professor in Computer Engineering at European University, Skopje, Macedonia.

TABLE OF CONTENTS

LIST OF CONTRIBUTORS

J. A. Castán Rocha
Faculty of Engineering, Autonomous University of Tamaulipas, Ciudad Victoria, Mexico

S. Ibarra Martínez
Faculty of Engineering, Autonomous University of Tamaulipas, Ciudad Victoria, Mexico

J. Laria Menchaca
Faculty of Engineering, Autonomous University of Tamaulipas, Ciudad Victoria, Mexico

J. D. Terán Villanueva
Faculty of Engineering, Autonomous University of Tamaulipas, Ciudad Victoria, Mexico

M. G. Treviño Berrones
Faculty of Engineering, Autonomous University of Tamaulipas, Ciudad Victoria, Mexico

J. Pérez Cobos
Faculty of Engineering, Autonomous University of Tamaulipas, Ciudad Victoria, Mexico

D. Uribe Agundis
TecNM/Instituto Tecnológico de La Laguna, Torreon, Mexico

Ossama E. Ramadan
Department of Civil, Construction, and Environmental Engineering, University of Alabama at Birmingham, Birmingham, USA

Virginia P. Sisiopiku
Department of Civil, Construction, and Environmental Engineering, University of Alabama at Birmingham, Birmingham, USA

Rajendra S. Parmar
Gujarat Technology University (GTU), Ahmedabad, India

Bhushan H. Trivedi
GLS Institute of Computer Technology, Ahmedabad, India

Olumuyiwa Samson Aderinola
Department of Civil and Environmental Engineering, Federal University of Technology, Akure, Nigeria

David Oladotun Owolabi
Department of Civil and Environmental Engineering, Federal University of Technology, Akure, Nigeria

Mohammad Mahdi Rezapour Mashhadi
Department of Civil & Architectural Engineering, University of Wyoming Office: EN 3084 1000 E University Ave, Dept. 3295 Laramie, WY 82071, USA

Shaun S. Wulff
Department of Statistics, University of Wyoming, Ross Hall 336, Laramie, USA

Khaled Ksaibati
Wyoming Technology Transfer Center 1000 E. University Avenue Department, 3295 Laramie, WY 82071, USA

Mahdi Rezapour
Department of Civil & Architectural Engineering University of Wyoming Office: EN 3084 1000 E University Ave, Dept. 3295 Laramie, WY 82071, USA

Azad Abdulhafedh
University of Missouri-Columbia, MO, USA

Ranjit Prasad Godavarthy
Upper Great Plains Transportation Institute, North Dakota State University, Fargo, North Dakota, USA

Eugene R. Russell
Department of Civil Engineering, Kansas State University, Manhattan, Kansas, USA

Xinxin Tang
College of Science, Guilin University of Technology, Guilin, China

Guangming Deng
College of Science, Guilin University of Technology, Guilin, China
Institute of Applied Statistics, Guilin University of Technology, Guilin, China

Kazuyuki Watari
Faculty of Environmental Engineering, The University of Kitakyushu, Kitakyushu, Japan
Geo Cluster Co. LTD, Japan

Weijun Gao
Faculty of Environmental Engineering, The University of Kitakyushu, Kitakyushu, Japan

Mohammed Aljoufie
Department of Urban and Regional Planning, Faculty of Environmental Design, King Abdulaziz University, Jeddah, Saudi Arabia

Nir Sharav
Transportation Research Institute, Technion, Haifa, Israel

Yoram Shiftan
Transportation Research Institute, Technion, Haifa, Israel

Hilda Karlsson
Industrial and Financial Management & Logistics, Department of Business Administration, School of Business, Economics and Law at University of Gothenburg, Göteborg, Sweden

Rickard Bergqvist
Industrial and Financial Management & Logistics, Department of Business Administration, School of Business, Economics and Law at University of Gothenburg, Göteborg, Sweden

Lars Brigelius
Industrial and Financial Management & Logistics, Department of Business Administration, School of Business, Economics and Law at University of Gothenburg, Göteborg, Sweden

Jesús Murgoitio
Industry & Transport Division, Tecnalia, Derio, Spain

Rafael Durbán
Grupo Avanza, Madrid, Spain

Joshué M. Pérez
Industry & Transport Division, Tecnalia, Derio, Spain

Antonio García
Grupo Avanza, Madrid, Spain

Jose A. Moreno
Grupo Avanza, Madrid, Spain

Ray A. Lattarulo
Industry & Transport Division, Tecnalia, Derio, Spain

Erick P. Massami
Unit for Applied Mathematics, Dar es Salaam Institute of Technology, Dar es Salaam, Tanzania

Benitha M. Myamba
Department of Logistics and Transport Studies, National Institute of Transport, Dar es Salaam, Tanzania

Leticia Edward
Department of Computing and Communication Technologies, National Institute of Transport, Dar es Salaam, Tanzania

Bomboma Kalgora
School of Economics and Management, Shanghai Maritime University, Shanghai, China

Ying Liu
College of Metropolitan Transportation, Beijing University of Technology, Beijing, China

Xiaoyi Li
Energy & Environment Center, Beijing Transport Institute, Beijing, China

Jun Yang
Beijing Jiaotong University, Beijing, China
Beijing Transport Institute, Beijing, China

LIST OF ABBREVIATIONS

AADT	Annual Average Daily Traffic
AADT	Average Annual Daily Traffic
ACC	Adaptive Cruise Control
AHP	Analytical Hierarchy Process
ANN	Artificial Neural Networks
BOOT	Build-Own-Operate-Transfer
CAPS	Center for Advanced Public Safety
CARE	Critical Analysis Reporting Environment
CBA	Cost Benefit Analysis
CBD	Central Business District
CDS	Crashworthiness Data System
CMU	Carnegie Mellon University
CMV	Commercial Motor Vehicle
CODES	Crash Outcome Data Evaluation System
CP	Crash Predictor
CRFs	Crash Reduction Factors
CWT	Cumulative waiting time
DBF	Database Files
DMUs	Decision Making Units
DOT	Department of Transportation
DSS	decision support system
DTA	Dynamic Traffic Assignment
DUI	Driving Under The Influence
EDRs	Event Data Recorders
EGNOS	European Geostationary Navigation Overlay Service
EMS	Emergency Service Costs
EOT	Equip-Operate-Transfer

ETD	European Transport Directive
ETS	Emission Trading Scheme
FCD	Floating Car Data
FEM	Fuzzy Entropy Method
FHWA	Federal Highway Administration
FMCSA	Federal Motor Carrier Safety Administration
FST	Fuzzy Set Theory
GDP	Gross Domestic Product
GES	General Estimates System
GIS	Geographic Information Systems
GPS	Global Positioning System
GTV	Global Traffic Volume
HD	High Definition
HOS	Hour of Service
HRRRP	High-Risk Rural Road Program
HSIP	Highway Safety Improvement Program
ITCS	Intelligent Traffic Control System
ITE	Institute of Transportation Engineers
ITLC	intelligent Traffic Light Controllers
KCARS	Kansas Crash Analysis Reporting System
KIT	Karlsruhe Institute of Technology
KMRT	Kaohsiung Mass Rapid Transit
LKS	Lane Keeping System
LLCs	landlocked countries
LOS	Level of Service
LTCCS	Large Crash Causation Study
MOEs	Measures of Effectiveness
MSAS	Multi-Functional Satellite Augmentation System
MSHP	Missouri State Highway Patrol
MSM	Modal Split Model
MUTCD	Manual on Uniform Traffic Control Devices
MVMT	Million Vehicle Mile Traveled
NASS	National Automotive Sampling System
NCSA	National Center for Statistics and Analysis

NHTSA	National Highway Traffic Safety Administration
NIS	Negative Ideal Solution
PDO	Property Damage Only
PIS	Positive Ideal Solution
PNDs	Personal Navigation Devices
PPP	Public-Private-Partnership
RAHCO	Reli Assets Holding Company
RPT	Road Public Transport
SAS	Statistical Analysis System
SDS	State Data System
SEM	Structural Equation Model
STARS	Statewide Traffic Accident Records System
TFN	Triangular Fuzzy Number
TOPSIS	Technique for Order Preference by Similarity to Ideal Solution
TRL	Tanzania Railway Limited
URT	Urban Rail Transport
UTOPIA	Urban Traffic Optimization by Integrated Automation
VMT	Vehicle Miles Travelled
VSL	Variable Speed Limits
WAAS	Wide Area Augmentation System
WHP	Wyoming Highway Patrol
WYDOT	Wyoming Department of Transportation
ZINB	Zero-Inflated Negative Binomial
ZIP	Zero-Inflated Poisson

PREFACE

In the last two decades, a significant revolution has taken place in all spheres of human life, caused by the explosive development of computer and information technology. Advanced information technology opens new horizons and develops new approaches to solving traffic problems. These new opportunities must (of course) be elaborated in the literature as a need for 21st century transport engineers.

On a global scale, among the expert circles in the field of transport, there is growing discussion regarding the discrepancies among the transport costs for the individual user and for the community. The notion of external transport costs, and the tendency to convert those costs from indirect to direct, are directions that will have significant impact in traffic transport engineering.

In the future, the progressive introduction of intelligent transport systems covering advanced traffic management systems, driver information systems, automated vehicle control - are expected. Particular attention will be given to the automatic toll collection, toll collection in city traffic, and the traffic restriction techniques.

Human civilization increasingly understands the need to protect the human environment. These efforts are expressed by the concept of "sustainable development". Traffic, as one of the major polluters of the environment, is certainly an area where the concept of sustainable development must be embedded. Of course, the education of modern traffic engineers must include new approaches and methods based on the concept of sustainable development of the transportation system.

This edition covers different topics from the transportation technologies, including control and automation of transport, public transport prediction, techniques to reduce crashes and case studies from different countries.

Section 1 focuses on control and automation of the transport, describing fuzzy rules to improve traffic light decisions in urban roads; evaluation of merge control strategies at interstate work zones under peak and off-peak traffic conditions; modulating traffic signal phases to realize real-time traffic control system; traffic regulation at critical intersections: a case study of Odole intersection, Akure, Nigeria.

Section 2 focuses on techniques to reduce crashes, describing comprehensive study of single and multiple truck crashes using violation and crash data; predicting truck at-fault crashes using crash and traffic offence data; road traffic crash data: an overview on sources, problems, and collection methods; road crash prediction models: different statistical modeling approaches; high-risk rural road safety study and determining the crash-reduction factors for high-risk rural-road usage.

Section 3 focuses on public transport prediction, describing prediction of civil aviation passenger transportation based on ARIMA model; present assessment of public traffic system based on GIS in Kitakyushu; development of a GIS based public transport composite social need index in Jeddah; evaluation of past investment in urban public transportation.

Section 4 focuses on case studies from different countries, describing transport policy - the perspective of Swedish stakeholders; Spanish initiative for the automation in urban transport: AutoMOST; fuzzy analysis and evaluation of public transport service quality: a case study of Dar-es Salaam city, Tanzania; intermodal terminal localization using a linear programming approach: the case study of Togo and west African landlocked countries; study on objectives and strategies for energy conservation and emission reduction in transportation industry during the 13th five-year plan period in Beijing.

SECTION I
CONTROL AND AUTOMATION OF THE TRANSPORT SOLUTIONS

CHAPTER

1

Fuzzy Rules to Improve Traffic Light Decisions in Urban Roads

J. A. Castán Rocha[1], S. Ibarra Martínez[1], J. Laria Menchaca[1], J. D. Terán Villanueva[1], M. G. Treviño Berrones[1], J. Pérez Cobos[1], D. Uribe Agundis[2]

[1]Faculty of Engineering, Autonomous University of Tamaulipas, Ciudad Victoria, Mexico

[2]TecNM/Instituto Tecnológico de La Laguna, Torreon, Mexico

ABSTRACT

Many researchers around the world are looking for developing techniques or technologies that cover traditional and recent constraints in urban traffic control. Normally, such traffic devices are facing with a large scale of input

Citation: A. Castán Rocha, J. , Ibarra Martínez, S. , Laria Menchaca, J. , D. Terán Villanueva, J. , G. Treviño Berrones, M. , Pérez Cobos, J. and Uribe Agundis, D. (2018), "Fuzzy Rules to Improve Traffic Light Decisions in Urban Roads". *Journal of Intelligent Learning Systems and Applications*, **10**, 36-45. doi: 10.4236/jilsa.2018.102003.

data when they must to response in a reliable, suitable and fast way. Because of such statement, the paper is devoted to introduce a proposal for enhancing the traffic light decisions. The principal goal is that a semaphore can provide a correct and fluent vehicular mobility. However, the traditional semaphore operative ways are outdated. We present in a previous contribution the development of a methodology capable of improving the vehicular mobility by proposing a new green light interval based on road conditions with a CBR approach. However, this proposal should include whether it is needed to modify such light duration. To do this, the paper proposes the adaptation of a fuzzy inference system helping to decide when the semaphore should try to fix the green light interval according to specific road requirements. Some experiments are conducted in a simulated environment to evaluate the pertinence of implementing a decision-making before the CBR methodology. For example, using a fuzzy inference approach the decisions of the system improve almost 18% in a set of 10,000 experiments. Finally, some conclusions are drawn to emphasize the benefits of including this technique in a methodology to implement intelligent semaphores.

Keywords: Fuzzy Inference System, Urban Traffic Control, Vehicular Mobility, Intelligent Transport System

INTRODUCTION

One of the most pursuit aspects to be improved in modern cities is its transit infrastructures. A traffic infrastructure is a dynamic and open scenario shaped by different kind of information which is used to make trustworthy and reliable decisions. In Mexico, the traditional way of controlling a junction is by using a traffic light signal. Such control was functional and suitable for the past. The number of cars and other relevant factors that affect the uses of the road are representing a technological challenge for researchers. In previous work [1] presented a methodology that makes an evaluation of the green light time based on the volume of cars to modify the duration of such light in the next phase of the semaphore. However, it was observed that sometimes the change in the green light duration was made and this change was not really significant to increase the level of service of the road. As a first conclusion, the fact of incurring in a non-adequate recommendation could cause both noise in the previously synchronized phase of the junction and increase the computational cost associated to the procedure of the system. The relevance of improving the way of adjusting the traffic light signals is quite needed due to current road constraints. For that reason, the paper proposes to include a prior review of some road aspects in order to decide if a modification of the green light interval could improve the service level of the junction. A fuzzy inference system is evaluated in order to corroborate the above statement.

A Fuzzy inference system can define whether semaphore must modify the time assigned to its green light with a high level of reliability. In fact, this technique is widely implemented to provide decisions under uncertain information. For instance, the fuzzy logic is a completely probabilistic technique defined by [2] as an extension of the traditional logic that uses concepts of ownership of an item which allows to measure the degree of membership of this element within a particular class and that has as main advantage the using linguistic terms to express knowledge, just as the human do in a natural way.

In [3] , the author grounds its arguments about safety in construction industries on a proper and opportune occupational risk assessment. Recent research concludes that such evaluation is quite difficult due to inadequate or uncertain data. For thus, the authors propose a Takani-Sugeno fuzzy inference systems to cover information lacks. The risk factor and controlling factors for accidental injuries are considered as input parameters. Analytical hierarchy process and subtractive clustering techniques are used both to evaluate safety levels of each type of injury and to reduce the number of rules in the ANFIS, respectively. To the end, a hybrid learning process is implemented to prove all the parameters corresponding to the input variables. The methodology has been applied to few selected construction sites in India.

Moreover, in [4] for the detection of fire use a diffuse model able to discern the areas that present a fire situation by recognition of colors. In this approach, the system generates a fuzzy search to locate fire, which handles processes only to the classification of binary map of the pixels of the background of the image. The fuzzy inference system follows a Mamdani process and consists of four stages: defuzzification of the variable, evaluation rules, composition and defuzzification, all based on the trapezoidal membership functions. The results show an improvement of almost 87% in comparison with the Bo-Ho Cho and Celik algorithms.

References [5] is proposed a fuzzy rule-based algorithm to evaluate job-satisfaction in an organization. The authors suggest interviews to obtain information about the relationship between effective factors and jobs facets. In particular, the Minnesota Satisfaction Questionnaire is used to obtain values of basis determinants of the respondents. Some fuzzy rules were generated from this information in order to compute the value of job satisfaction using the expert system shell ESPLAN. The proposed approach shows relevant advantages to operate with imperfect information for the evaluation of job satisfaction by using fuzzy logic.

Besides, in [6] a study to show how to put an expert system in practice to organize the security screening process is proposed. Authors highlight the relevance of the expert's opinion and problems involved in the imprecise nature of information to take advantages of fuzzy sets and fuzzy inference system to generate FUPSCA (Fuzzy Passenger Security Control Assessment). It is used to set the parameters of different variables related to the current level of terrorist threat in airports. In conclusion, the paper argues that in a near future, the use of multi-criteria evaluation such as capacity, effectiveness or passenger comfort to manage passenger security control systems will be necessary.

All the aforementioned approaches are evidence of many works oriented to provide effective approaches to solve well-detected lacks in transport area. The main conclusion of these works aims to continue the evaluation of studies on three specific subjects: 1) the interaction among autonomous semaphores; 2) the capability of the system to be robust under uncertain data; and 3) the reliability on the semaphore decisions. About such statement, the paper argues that the implementation of the support decision technique studied appears to be a suitable and robust way to decide when a semaphore must attempt to change the duration of a green light. Different variables can be implicated in this important and dynamic decision. However, FIS method arises as a proper solution to be a filter which avoids that semaphore performs unnecessary actions. Therefore, the motivation of this paper is to strengthen the operatively of a novel methodology for controlling semaphores in an autonomous way using a CBR approach. The remainder of this paper is organized as follows: Section 2 introduces the specific aspects of our proposal. Section 3 presents the results obtained in the experiments. Finally, Section 4 emphasizes the main contributions of the paper and defines some ideas for further progress on this interesting domain.

FUZZY INFERENCE SYSTEM IN TRAFFIC COTROL

Modern cities are looking for offering a suitable traffic service level from different perspectives (i.e., urbanism, town planning or computer science). The idea is not only to increase the productiveness of the city but also it is to reach a more comfortable, secure and healthy environment for all the citizens. Among all the particularities of this utopia, roads better administered are one of the most studied topics in the actuality. In this sense, the authors present in [1] , a general methodology capable of evaluating some aspects on the road in order to modify the green light interval of a semaphore using a CBR approach. However, one data among the results demonstrate that in occasions modifying such interval is not the appropriate response of the system.

For example, in some cases, the system recommends aggregating 2 or 3 seconds to the green light interval and this action does not reflect an increment in the service level of the analyzed line. In spite of this, it is necessary to evaluate real information related to the road conditions before of initializing the processes of the methodology. To do this, the authors propose to include a FIS technique to define if it is suitable to do a modification in the green light duration of a determined semaphore. This method will allow the system to validate a set of attributes to determine, prior to the start of a green light, if according to the conditions of the road, it will be necessary modify the duration of such light.

To consider a possible change of the green light duration in an isolated 4-lane traffic intersection using a FIS (Fuzzy Inference System), we propose a Mamdani model with five input variables such as:

- Number of incoming cars in each line of this specific road.
- Green light time assigned to the semaphore in its previous intervention in the junction phase.
- Day time in real clock scheduling.
- Type of road that semaphore should administrate.
- Traffic condition at the moment of the FIS decision.

Each of these input variables are represented by a trapezoidal membership function. The response of the model (i.e., determine whether or not to change the duration of the green light) is then characterized by a triangular function as is illustrated in Figure 1. For sake of simplicity, the proposed approach defines a set of attributes for each input variable which are introduced in Table 1. These values were selected and calculated based on previous work presented in [1] using experience of human experts on traffic control.

In this way, the graphical representation of the membership functions of the linguistic attributes for each of the defined variables, by using the

MatLab diffuse control is presented in Figure 2.

In order to ensure a proper application of green light times in a semaphore, the output variable called decision is represented by a dichotomous response (YES or NO) (Figure 3). To note, this new system procedure only is in-charge to decide if the semaphore should evaluate whether the green light interval must be modified. Only two answers are defined: “No”, which means that according to the FIS, the information provided for the road is not significant enough to indicate that a change in the duration of the green light should reflect a better vehicular mobility. Contrary to this, the “Yes” response represents the membership of the road conditions in which is recommended to do a change in the time of the green light. To express the above, a set of 56 rules was generated, where each variable was combined with the rest, so that all possible combinations were considered, as illustrated in Figure 4 (only 30 rules are presented).

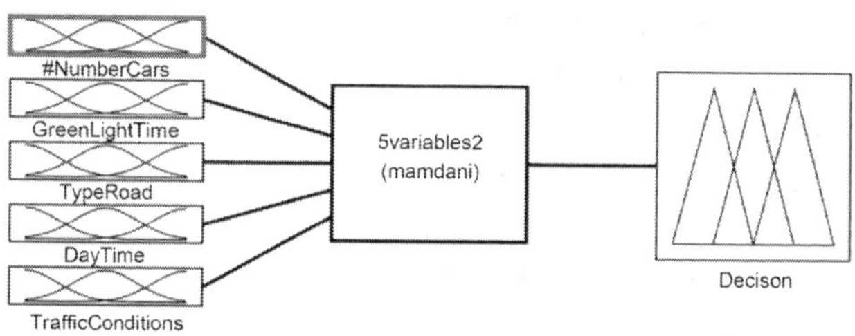

Figure 1. Scheme of the Mamdani model to evaluate semaphore decisions.

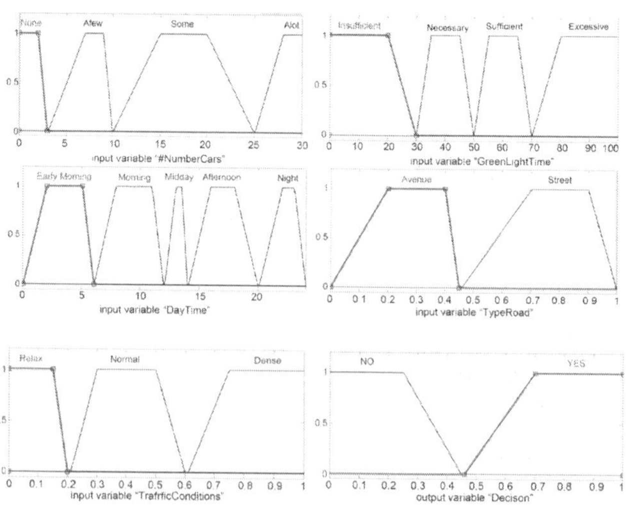

Figure 2. Membership representation of the attributes for each input variable and decision.

EXPERIMENTAL RESULTS

All the experiments were developed on a Macbook Pro with the following specification:

- macOS Sierra OS10.12.6 version,
- Intel Core i7 2.90 GHz Processor,
- 8 GB 1600 MHz DDR3 memory,
- 1 TB SSD storage.

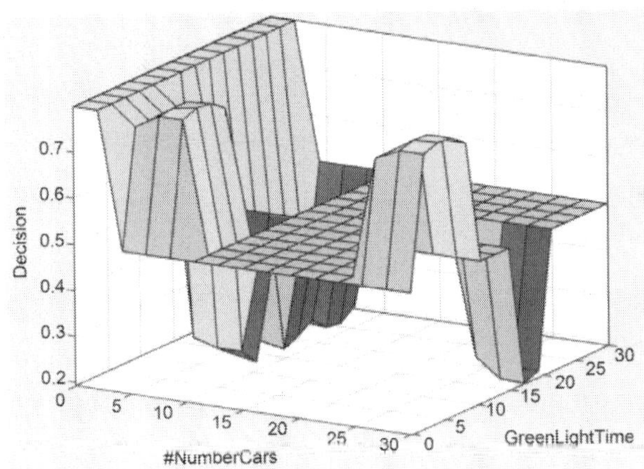

Figure 3. Surface view for number of cars and green light time.

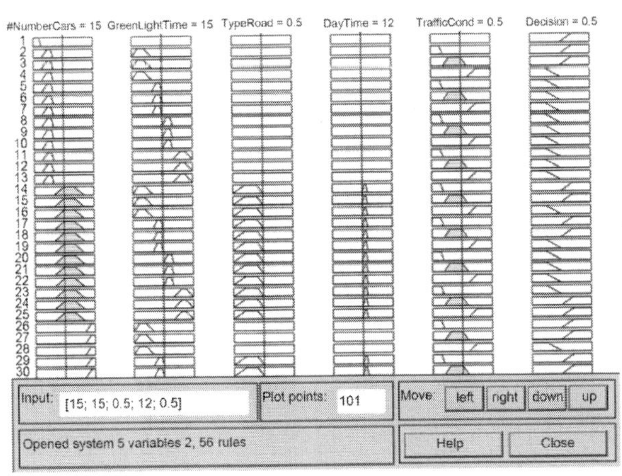

Figure 4. Review of the behavior rules of the diffuse inference system based on the Mamdani model.

To validate the efficiency of the generated FIS, a phase of 10,000 control experiments were performed each, taking as reference the representation of an infrastructure of four roads, one of the most used in Mexico (see Figure 5) with the following characteristics:

- Each road has movements in both directions.

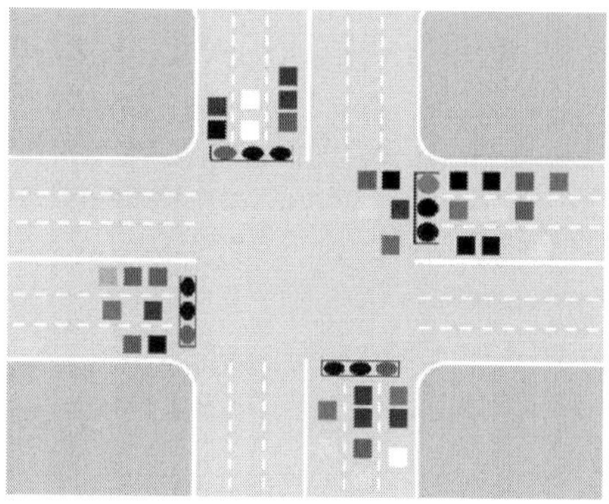

Figure 5. The representation of an infrastructure of four roads used in the test.

Table 1. Relation variable attributes for the fuzzy decision model.

VARIABLE	ATTRIBUTE	VALUE
Number of Cars	None	0 - 3
	A Few	3 - 10
	Some	10 - 25
	A lot	25 or more
Green Light Time	Insufficient	0 - 30
	Necessary	30 - 50
	Sufficient	50 - 70
	Excessive	70 - 100
Day Time	Early Mourning	0 - 6
	Mourning	6 - 12
	Midday	12 - 14
	Afternoon	14 - 20
	Night	20 - 24
Type of Road	Avenue	0.00 - 0.45
	Street	0.45 - 1.00

Traffic Conditions	Relax	0.00 - 0.20
	Normal	0.21 - 0.60
	Dense	0.61 - 1.00

- It can turn left (according to the traffic laws that are regulated in Mexico).
- Continuous right turn (i.e., not necessarily the semaphore should have green light, simply the driver must be cautious).
- Each road has three lanes.
- Average speed of the car 50 km/h.
- Different types of cars (i.e., sizes and shapes).

The results are compared against the performance obtained from the system when it modifies the green light time at all turns of the semaphore cycle. In this sense, if the system uses a decision support technique, such as the FIS technique, its performance increases by almost 18% as shown in Figure 6.

Another interesting point of view to be noted is the number of experiments each one of the proposed approaches needed to reach stability in the overall system performance. In other words, this data reflects the number of situations that system requires to get sufficient information into their data base in order to reach trustworthy and reliable decisions. For instance, in Figure 7 the computed curve using a sliding window up to the current trial with a smooth order 3 method for the performance of the system under different solution perspectives is presented. As can be noted, the system requires almost 1000 cases to reach a more stable behavior in order to ensure a suitable and constant mobility level.

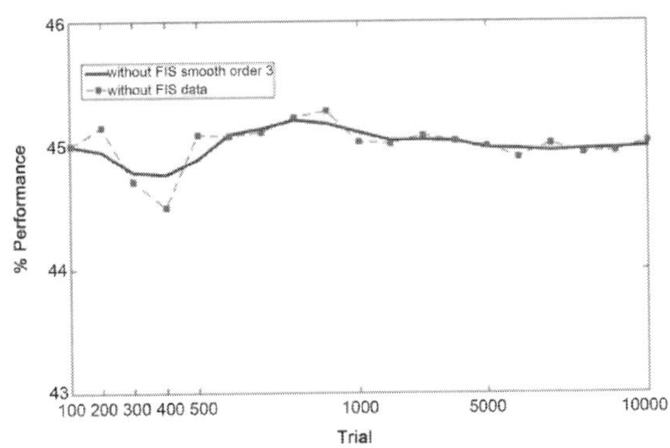

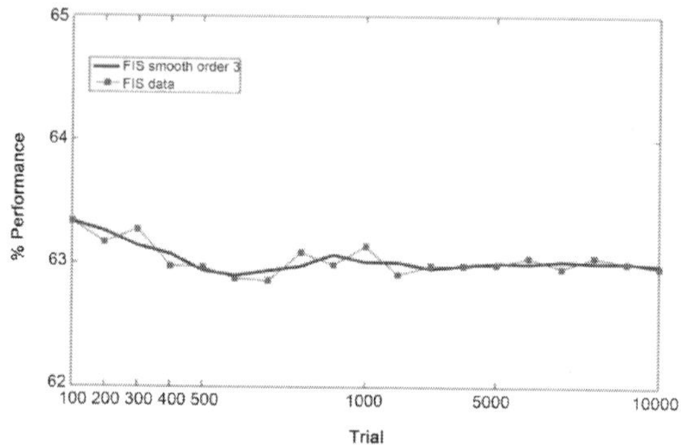

Figure 6. Performance obtained when the system uses a FIS to decide whether or not to modify the green light time.

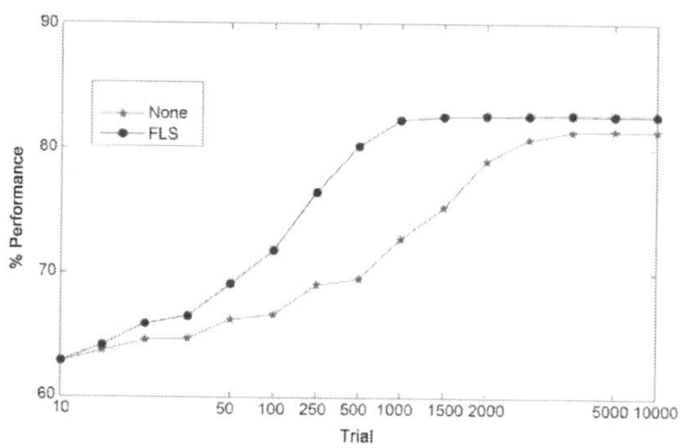

Figure 7. Comparison of the computed curve of trial experiments to reach system stability.

CONCLUSIONS AND FUTURE WORK

Road traffic congestion is one of the main causes in low productivity and the decrease in modern city standards. In this sense, some recent trends in artificial intelligence suggest that in the near future, intelligent agents will improve some road challenges. In this light, the present paper introduces the use of a fuzzy inference system to define when it will be opportune to change the length of the green light in a determine traffic light signal looking for improving the level of service offered in a particular junction.

The results show that using a decision support technique, such as a fuzzy logic rules, the system performance increases in almost 18%. Therefore, it is possible to identify that the use of this technique allows the system to modify the duration of the green light, improving its decision-making process which is reflected in an improvement of its operative capacity, its reliability and certainty. It is important to remark that for experimental tests, the human factor had a vital role because an expert in traffic control recommends the variables and the rules consider in the FIS. For future approaches, it is necessary considered to expand the study about which data should be taken into account in order to reach a suitable and trustworthy set of real variables.

From the obtained results, it is possible to argue that the proposed approach achieves significant benefits in vehicular mobility, in terms of congestions by increasing the traffic capacity of any intersections. However, future investigation is suggested including the following aspects: "Issues such as acceleration and des-acceleration, collision avoidance and vehicles with different velocities, must be included in the simulator by offering a set of more realistic situations in the virtual instrument proposed here".

Indeed, people are different and react modifying their behavior pattern, considering variables that have not impact on the situation. Drivers are the most important source of data. At the moment, the literature has not reported a system robust enough to satisfy the road constraints of the cities. Intelligent transport systems must be closely designed according to the human beings. In this light, it is necessary to develop inference models based on humans where emotions, beliefs, politics, manners, etc., can be considered. In this light, it is necessary to develop inference models based on human must consider emotions, beliefs, politics, manners, etc.

REFERENCES

1. Ibarra, S., Castán, J.A. and Laria, J. (2014) Optimizaing Urban Traffic Control Using a Rational Agent. Journal of Zhejiang University Science C, 15, 1123-1137. https://doi.org/10.1631/jzus.C140003
2. Zadeh, L.A. (1965) Fuzzy Sets. Information and Control, 8, 338-353. https://doi.org/10.1016/S0019-9958(65)90241-X
3. Debnath, J., Biswas, A., Sivan, P., Sen, K.N. and Sahu, S. (2016) Fuzzy Inference Model for Assessing Occupational Risks in Construction Sites. International Journal of Industrial Ergonomics, 55, 114-128. https://doi.org/10.1016/j.ergon.2016.08.004
4. Buemi, A., Giacalone, D., Niccari, F. and Spampinato, G. (2016) Efficient Fire Detection Using Fuzzy Logic. 6th International Conference on Consumer Electronics, Berlin, 5-7 September 2016, 47.
5. Saner, T., Gardashova, L., Allahverdiyev, R. and Eyupoglu, S. (2016) Analysis of the Job Satisfaction Index Problem by Using Fuzzy Inference. Procedia Computer Science, 102, 45-50. https://doi.org/10.1016/j.procs.2016.09.368
6. Skoruoski, J. and Uchronski (2016) Managing the Process of Passenger Security Control at an Airport Using the Fuzzy Inference System. Expert Systems with Applications, 54, 284-293. https://doi.org/10.1016/j.eswa.2015.11.014

CHAPTER 2

Evaluation of Merge Control Strategies at Interstate Work Zones Under Peak and Off-Peak Traffic Conditions

Ossama E. Ramadan, Virginia P. Sisiopiku

Department of Civil, Construction, and Environmental Engineering, University of Alabama at Birmingham, Birmingham, USA

ABSTRACT

Maintenance and rehabilitation projects of interstate facilities typically mandate lane closures. Lane closures require merging maneuvers that often result in reduced speeds and traffic bottlenecks. Work zone impacts on traffic operations are magnified when project durations are extended. Conventionally, work zone traffic control plans are developed to address

Citation: Ramadan, O. and Sisiopiku, V. (2016), "Evaluation of Merge Control Strategies at Interstate Work Zones under Peak and Off-Peak Traffic Conditions". *Journal of Transportation Technologies*, **6**, 118-130. doi: 10.4236/jtts.2016.63011.

work zone impacts. This study evaluated various merge control strategies at interstate work zones peak and off-peak traffic conditions and summarized related impacts. A comprehensive microscopic simulation model was developed in full consideration of driver/vehicle behavior at work zones. The analysis of simulation results revealed that merge control strategies, when implemented during peak and off-peak conditions, can preserve the level of service and provide favorable mobility, safety, and environmental impacts. In addition, results indicate that transportation agencies' practice of scheduling work zone activities during the off-peak may not be the most optimum approach. Overall, the findings of this study highlight the need for evaluation of work zone scheduling practices in full consideration of agency, user, and project costs.

Keywords: Traffic Simulation, Bottlenecks, Work Zone Traffic Control, Traffic Congestion

INTRODUCTION

Highway maintenance and rehabilitation projects are essential for maintaining the nation's infrastructure functionality. Such projects often mandate lane closures to provide a work space, which in turn result in bottlenecks and disruption of traffic [1] . Studies indicate that work zones represent the second largest cause of nonrecurring delay on principal arterials with nearly 24% of all nonrecurring delay attributed to work zone operations [2] .

In an effort to reduce undesirable impacts from work zone presence, transportation agencies opt to schedule maintenance activities during off-peak hours. However, such practice may lengthen project duration and increase maintenance-related costs as a result of repetitive setup, frequent cessation, and associated costs to nighttime operations. This is worrisome, especially given the fact that transportation agencies are often pressured to reduce the time they spend on construction to minimize the impact on the traveling public and local businesses [3] .

Another common practice is the use of traffic control to facilitate vehicle movement upstream and through the work zone. Conventional traffic control plans normally work well when traffic demand is less than the capacity of the open lanes. However, when demand exceeds capacity, both traffic operations and safety at work zone locations are compromised, especially when the congestion extends upstream beyond the advance lane closure signs [4] .

A need exists to evaluate various merge control strategies at interstate work zones under peak and off-peak traffic conditions and determine optimal practices for scheduling of construction activities that mitigate bottleneck effects at work zones while also considering project costs and scheduling efficiency constraints.

Background

Transportation agencies rely on traffic control plans to minimize the traffic impacts from work zone presence. Most agencies currently follow the conventional temporary traffic control strategies set by Part 6 of the *Manual on Uniform Traffic Control Devices* (MUTCD) [5] . In an earlier study, the authors [6] performed a comprehensive literature review and a national survey of practices where they identified alternative traffic control strategies that hold promise toward improving effectiveness as compared to conventional merge control. Such alternative control strategies were categorized in that study as follows:

- Late merge control, which encourages drivers to retain their lanes until they reach the lane closure taper. Late merge is sometimes referred to as "zipper merge" when associated with an alternating "zipper" fashion merge maneuver into the open lane(s). Late merge control has been identified as the second most implemented bottleneck merge control strategy at interstate work zones.
- Early merge control, which encourages drivers to merge into open lane(s) sooner than they would do, by providing advance signs ahead of lane closure(s) or merge tapper(s). Early merge control has been identified as the most implemented bottleneck merge control strategy for interstate work zones.
- Mainline merge metering strategies, which are built upon the same architecture as late merge control; however, a merge meter, similar to a conventional ramp meter, is installed at the tapper or point of merge to regulate the release of vehicles trying to merge into open lanes. Literature review indicated that mainline merge metering is not implemented by transportation agencies in the US.
- Temporary ramp metering, which builds upon the well-established benefits of ramp metering as an effective strategy for improving traffic flow along the mainline by controlling the rate of vehicles entering from the on-ramp into the mainline traffic.

The documented mobility and productivity benefits of ramp meters motivated researchers to study the use of temporary ramp meters as an alternative to managing traffic on interstate work zones located near or at on-ramps. Literature review indicated that temporary ramp metering is not currently implemented by any transportation agency in the US.

Problem Statement and Research Objective

Earlier literature review and a national survey of agencies practices indicated that there are four alternative bottleneck merge control strategies for interstate work zones [6] . However, two of the identified strategies are still a matter of research, indicating a gap between research and practice. In their earlier literature review and survey of practices, the authors noticed that all existing studies evaluate these bottleneck merge control strategies only under off-peak conditions. Moreover, literature and agency practices indicate the lack of formal guidance on selection and assessment of relevant impacts for the identified strategies. In addition, available literature lacks a comprehensive comparison between these four strategies. Accordingly, there is a need to study and evaluate the identified bottleneck merge control strategies at work zones, to better understand their impacts and benefits, and to provide some guidance to agencies for selecting an appropriate strategy.

The objectives of this study are (a) to compare the previously identified four strategies at interstates work zones of various lengths under peak and non-peak traffic conditions, and (b) report their associated impacts on traffic operations and the environment. The stated objectives can be achieved by providing answers to the following questions: What are the impacts of each strategy? Which strategy provides the most benefits? And, is scheduling of work zone activities during off-peak periods justified?

METHODS

The focus of this study is to consistently capture detailed elements of traffic control within the immediate work zone area. According to the comprehensive work zone modeling and simulation study that was performed by Hardy and Wunderlich [7] , the most appropriate methodology would be a mono-scale method that utilizes microscopic simulation. Therefore, this study used of *Corridor Simulation* (CORSIM) version 6.3 for developing a stochastic microscopic simulation model of a section of I-65 in Birmingham, Alabama, as shown in Figure 1. This is the latest version of CORSIM that was released by McTrans Center at the University of Florida for the US *Federal Highway Administration* (FHWA). The study segment is approximately 14 miles (22.53 Km) in length and extends from exit 247, where it intersects Valleydale Road, to exit 261, where it intersects Interstate 20/59 (I-20/59). The study segment typically has three 12-ft (3.66 m) lanes per mainline direction, with auxiliary lanes added at ramps locations. The posted speed limit on this study interstate is generally 60 mph (96.6 km/h) with an advisory speed limit of 45 mph (72.45 km/h) on ramps. The 2014 *Annual Average Daily Traffic* (AADT) for this study section is 122,510 veh/day, with peak traffic typically occurring on weekdays between 7:00 am and 9:00 am, and off-peak traffic typically occurring between 2:00 am and 4:00 am [8] . The following subsections illustrate base model development, base model calibration, setting up the microsimulation platform for work zone simulation, and simulation of each of the identified four merge control strategies.

Base Model Development

Two base models were developed to simulate typical existing, or baseline, peak and off-peak traffic conditions. To develop these base models, data related to road geometry, pavement conditions, traffic volumes, and traffic characteristics were collected. These data were collected during the period February 2014 to February 2015 at the beginning of the study segment, at each entry and exit ramp, at mid-sections between ramps or where geometry would change, and at the end of the study segment. This study followed the guidelines and methods stated by the FHWA Traffic Analysis Tool Box, Volumes III, IX, and IX [7] [9] [10] and the *Institute of Transportation Engineers* (ITE) Manual of Transportation Engineering Studies [11] to setup and calibrate CORSIM. To check the quality of the base models, a review of physical inputs and traffic volumes was conducted to identify any errors that may have occurred while developing the base models. The review did not identify any errors or non-conformities in the physical or traffic inputs including acceleration/deceleration lane lengths; number of lanes and lane alignment; lane drops/lane adds; free flow speeds; curvature; grades; freeway volumes for each time period balance; and overall freeway volumes balance.

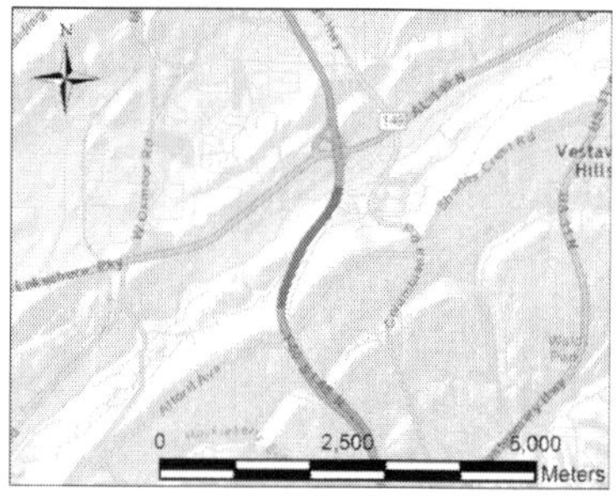

Figure 1. Map showing study area.

Base Model Validation and Calibration

CORSIM has been extensively validated by several studies over the past 45 years [10] . Earlier studies provide confidence in the overall model ability to predict traffic performance such as traffic volumes, travel times, average speeds, and average delays. However, any base model has to be calibrated for local conditions so that the simulated volumes and speeds would be within of 15% of actual volumes and speeds [10] . Calibration data composing of average speeds and link volumes were collected within uniform sections along the study segment during the period February 2014 to February 2015. Speed data were collected by test vehicle method equipped with after-market GPS units. Volume data were obtained by manual observation using a handheld electronic count board. Simulated volumes and speeds were compared against field collected volumes and speeds, relative errors were calculated using Equation (1), and were plotted for each simulation period comprising 15 minutes interval. The calibrated model produced traffic volumes and speeds within a range of ±10%, with slightly higher traffic volumes than those of the actual facility, and slightly lower speeds than actual speeds, as illustrated by Figure 2. The base model calibration yielded satisfactory results and provided confidence in the model's ability to replicate closely actual conditions.

$$\text{Relative Error} = \frac{\text{Simulated Data} - \text{Measured Data}}{\text{Measured Data}} \times 100 \quad (1)$$

Because CORSIM is a stochastic time-step microsimulation model, the *Output Processor* was configured to use the same *random number seeds* across all simulation runs. In addition, mean cumulative outputs and corresponding standard deviations were collected from 42 scenarios that were run five times each with a total of 210 simulation runs. Manipulating CORSIM for simulating driving conditions in work zones has been extensively studied by other researchers, and seven parameters were identified along recommended values to setup CORSIM for work zone driving conditions independent of location or geometry [7] - [10] [12] [13] . These seven parameters are:

1). Vehicle entry headway: This is the method CORSIM will use to generate vehicles at entry nodes. A negative exponential distribution was used in this study, which can be achieved by setting CORSIM to use an Erlang Distribution with a = 1.

2). Rubberneck factor: The rubberneck factor, in a percentage, represents the reduction in capacity that results from the tendency of drivers at work zones who to slow down to see what is happening. Literature recommend that open lanes at work zones as well as a portion equal to work zone length of the closed lane(s) post work zones should be coded for rubber necking incidents with a factor of 20% to account for such phenomenon.

3). Car following sensitivity multiplier: CORSIM uses driver sensitivity to account for the car-following logic. Despite the fact that drivers tend to be more aggressive at work zones, literature recommends using the default car-following sensitivity multipliers.

4). Mean start-up delay time: This is a link-specific parameter that is used by the model to discharge vehicles from a ramp meter onto the freeway. The default value is 1.0 second, and literature recommends a value of 2.4 seconds when simulating work zone conditions.

5). Time to complete lane-change maneuver: This parameter specifies the time needed by a driver to complete a lane-change maneuver. The default value is 2.0 seconds, and a literature recommended using a value of 2.2 seconds when simulating work zones to account for the increased driver aggressiveness at work zones, which does not usually allow drivers to complete a lane change without getting some resistance from other drivers.

6). Percent drivers yielding to merging vehicles: This parameter specifies the percentage of drivers desiring to yield the right-of-way to lane-changing vehicles attempting to merge ahead of them. The default value is 20%, and literature recommended a value of 15% to account for work zone conditions.

7). Minimum separation for generation of vehicles: This parameter governs the maximum rate at which vehicles can be emitted onto the network. The default value of 1.6 seconds is appropriate for freeway operations; however, earlier studies proved that a value of 1.3 seconds will enhance the calibration of CORSIM for work zone conditions.

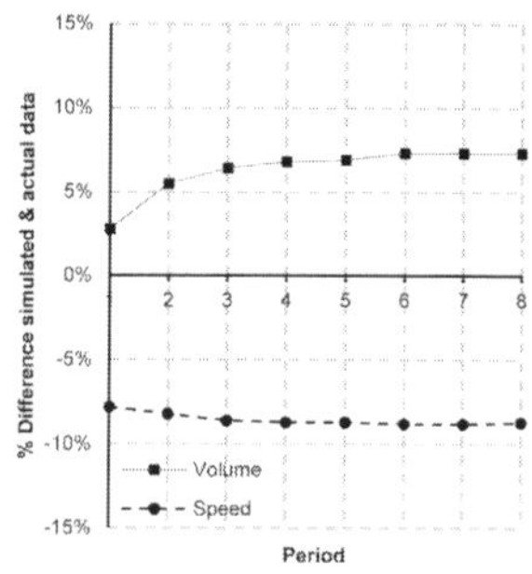

Figure 2. Speed and volume differentials for calibrated base model.

The following sub-sections explain how each strategy was modeled in CORSIM in the study herein.

Microscopic simulation of late merge control. Late merge control strategy depends on urging drivers to remain in the lane(s) that are about to be blocked until they reach the beginning of the tapper. This strategy attempts to consume all available capacity to accommodate as many vehicles as possible. Simulation was performed using the CORSIM built-in Incident Management Algorithm, for the aforementioned hypothetical work zone configurations. To account for the work zone presence, two incidents were coded. The first incident is a blockage on the rightmost lane starting from the beginning of simulation and extending to 2.5 hours to account for the warm-up time, and 20% rubbernecking-only on the middle lane and the left lane. The second incident is coded on the segment following the work zone and with the same length with 20% rubbernecking-only on the rightmost lane. Figure 3 provides a screen capture in CORSIM illustrating the microsimulation setup for late merge control.

Microscopic simulation of early merge control. Early merge control strategy depends on urging drivers to merge into open lane(s) as early as possible in a manner that minimizes lane changes near or at work zones. This strategy is typically used for lower volume facilities. Simulation was performed using CORSIM as a lane drop for the rightmost lane and a lane add following the work zone. In addition, rubbernecking-only incidents were coded along the work zone for open lanes to account for the speed reduction that occurs in work zones. Similar to late merge control, the model accounted for a buffer space; however, the merge taper was not accounted for as the model considered an early merge effective 5800 ft (1769.00 m) upstream of work zone. The distance of 5800 ft (1769.00 m) accounts for merging signs placed 800 ft (244.00 m) upstream of work zone and every 1000 ft (305.00 m) thereof. Figure 4 provides a screen capture in CORSIM illustrating the microsimulation setup for early merge control.

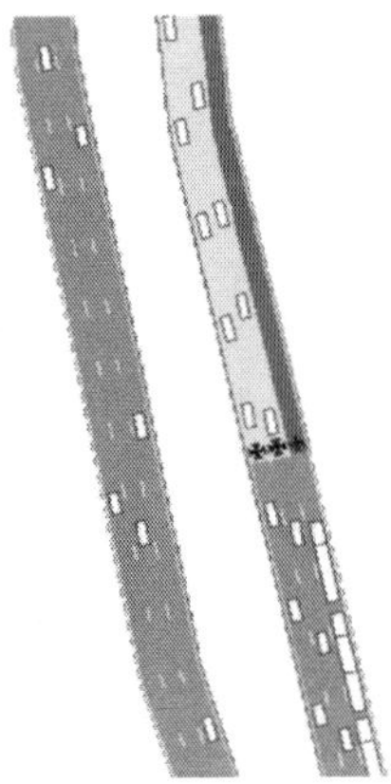

Figure 3. Work zone setup in CORSIM using late merge control.

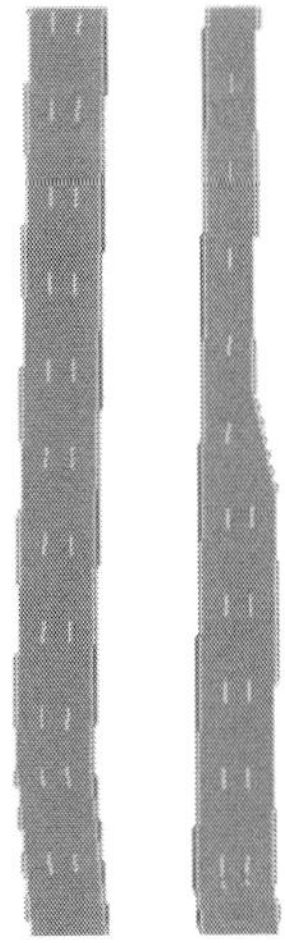

Figure 4. Work zone setup in CORSIM using early merge control.

Mainline merge metering microscopic simulation model. Mainline merge metering incorporates a merge meter that functions in the same manner as a ramp meter but is installed on the mainline of the freeway lane that is about to be dropped and vehicles have to merge in the adjacent open lane(s). The simulation model was coded to account for a mainline meter upstream of the work zone buffer zone, with an acceleration/merge lane of 300 ft (91.50 m) [14] . The configured meter was a demand/capacity meter with capacity of 2300 veh/hr/lane [15] [16] . The CORSIM demand/capacity metering algorithm performs an evaluation of current excess capacity, immediately downstream of the metered lane, at regular intervals, based on counts from the surveillance detectors on the freeway mainline [17] . A maximum metering rate is calculated such that the capacity of this freeway section is not violated. This calculated metering rate is then applied like clock-time metering. A minimum metering rate of three green signals/60 seconds is applied to ensure that waiting vehicles are not trapped between the meter and the ramp connection to the freeway. The metering rate is also limited to headways that are greater than two seconds. Figure 5 provides a screen capture in CORSIM illustrating the microsimulation setup for mainline merge metering.

Temporary ramp metering microscopic simulation model. Temporary ramp meters are installed at on-ramps where an on-ramp feeds an interstate work zone. The purpose of the temporary ramp meter is to regulate traffic going onto the interstate hence mitigating the mobility impacts of excessive merging in such sections. A hypothetical 3-to-2 work zone was modeled on the leftmost lane downstream from the subject on-ramp. A temporary ramp meter was installed with demand/capacity algorithm similar to the one modeled in the mainline metering strategy. Figure 6 provides a screen capture in CORSIM illustrating the microsimulation setup for temporary ramp metering.

RESULTS AND DISCUSSION

CORSIM has the capability to report several *Measures of Effectiveness* (MOEs) for the simulated interstate work zone configuration. These MOEs are either link-based or network-based. Link-based MOEs provide an understanding of the localized impacts of simulated merge control strategies at and in the vicinity of the work zone, while network-based MOEs capture the global impact on the simulated network. Selected MOEs were collected and averaged over five runs for each scenario while preserving the random seeds and are discussed in the following sections. Results were compared against preconstruction, or baseline conditions, conditions. Figures 7-13 illustrate the percentage change from preconstruction conditions for each MOE during peak and off- peak traffic. To simplify the comparison against preconstruction (baseline) conditions, an MOE Index (I_{MOE}) was calculated as illustrated by Equation (2), to represent MOEs compared to that of preconstruction conditions.

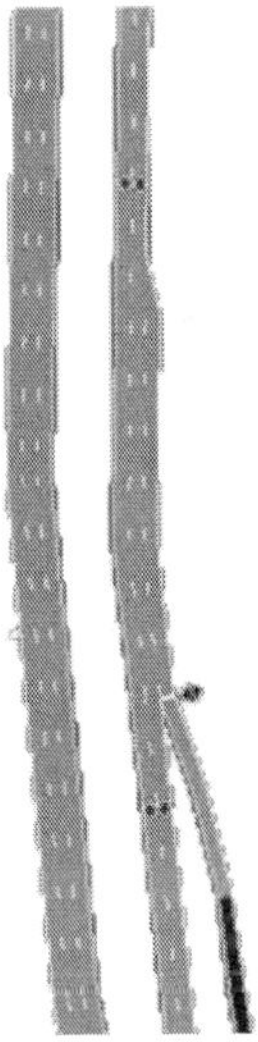

Figure 5. Work zone setup in CORSIM using mainline merge metering.

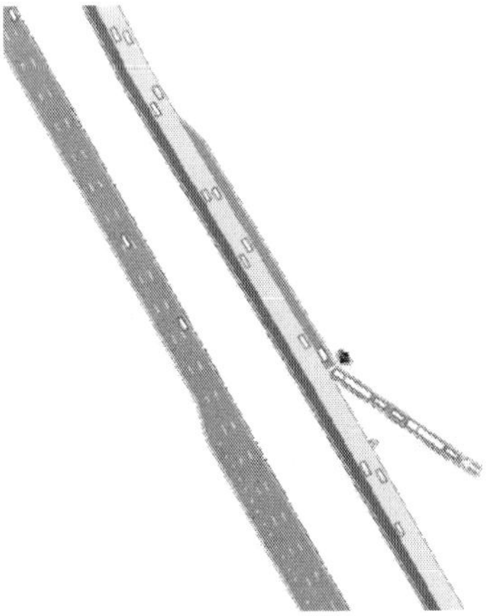

Figure 6. Work zone setup in CORSIM using temporary ramp metering.

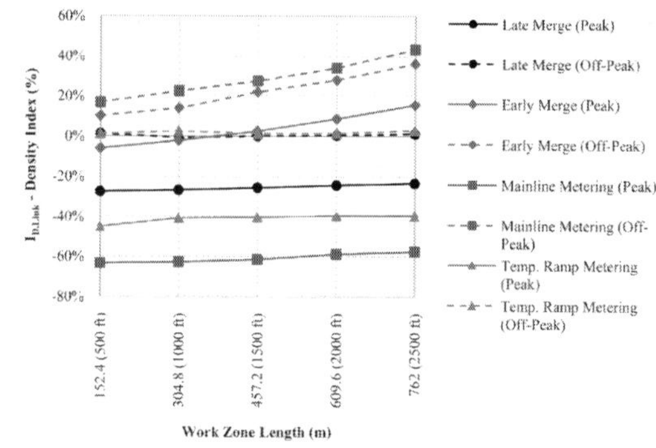

Figure 7. Density index for each merge control strategy by work zone length.

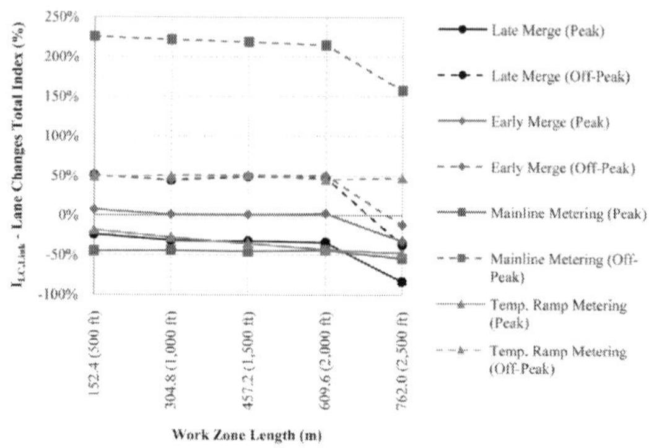

Figure 8. Lane changes index for each merge control strategy by work zone length.

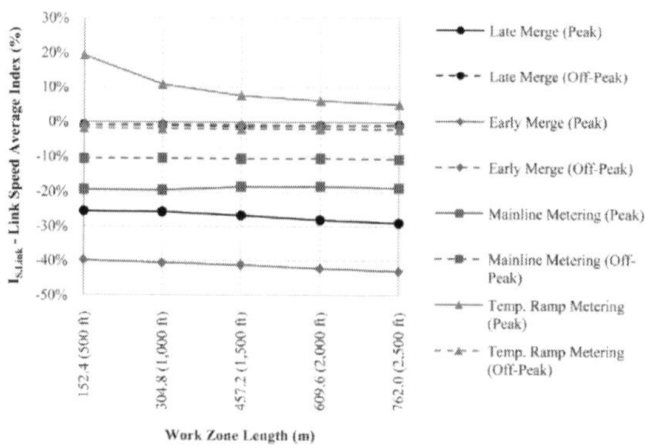

Figure 9. Link speed average index for each merge control strategy by work zone length.

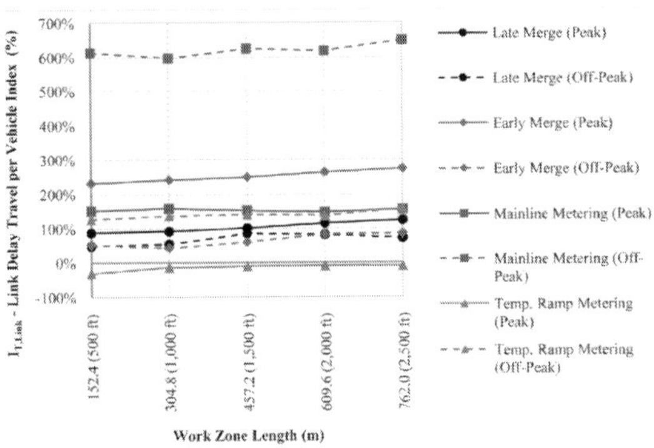

Figure 10. Link delay per vehicle index for each merge control strategy by work zone length.

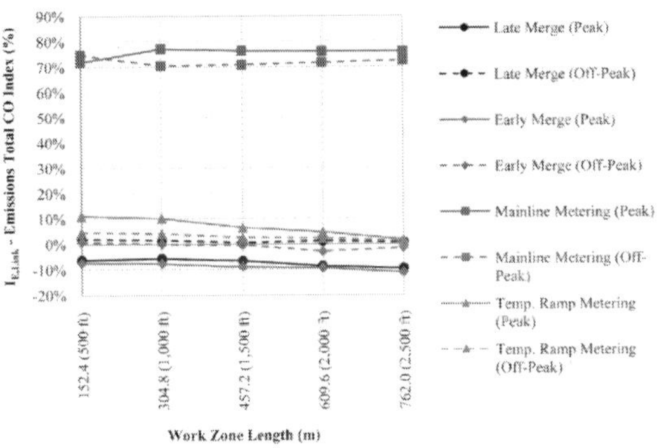

Figure 11. Link emissions total CO index for each merge control strategy by work zone length.

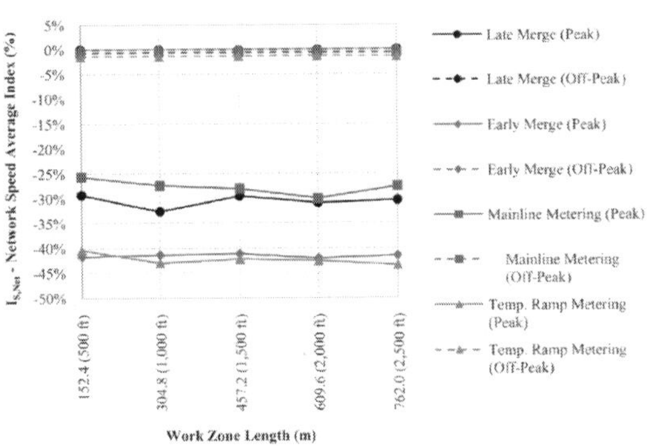

Figure 12. Network speed average index for each merge control strategy by work zone length.

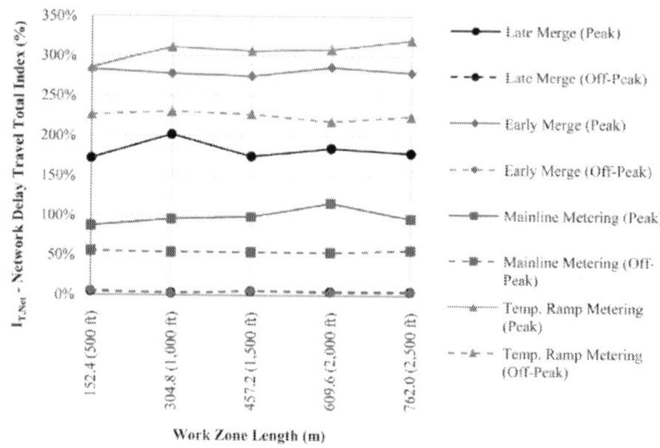

Figure 13. Network delay travel total index for each merge control strategy by work zone length.

$$I_{\mathrm{MOE},i,j} = \frac{\mathrm{MOE}_{i,j} - \mathrm{MOE}_{i,PC}}{\mathrm{MOE}_{i,PC}} \times 100 \qquad (2)$$

where, $\mathrm{MOE}_{i,PC}$ is the preconstruction MOE of the segment for a given traffic volume (i), and $\mathrm{MOE}_{i,j}$ is the same MOE for a given traffic volume (i) and a given merge control strategy (j). A positive I_{MOE} value indicates that the facility is experiencing higher levels of that MOE compared to preconstruction conditions, and similarly a negative I_{MOE} value implies lower levels than that of preconstruction conditions. Preconstruction conditions were chosen as baseline conditions since the purpose of implementing any of these merge control strategies is to preserve whatever conditions existed before having a work zone.

Link-Based MOEs

Density per lane (vehicles/Km/lane). Density is the key measure for *level of service* (LOS) on freeway segments. A density index ($I_{D,Link}$) was calculated, to represent density changes compared to that of preconstruction conditions. A positive ID,Link indicates that the facility is experiencing more density that preconstruction conditions, and negative ID,Link indicates improved density. Weighted average density was used following the guidelines of the Highway Capacity Manual [18] . The density index was plotted for each merge control strategy by work zone length. Figure 7 illustrates the percentage change of density per lane from corresponding preconstruction conditions by work zone length for both peak and off-peak traffic.

It is observed that under peak traffic conditions, late merge control, temporary ramp metering and mainline merge metering yielded lower densities compared to preconstruction conditions. However, early merge control yielded improvements with shorter work zones and higher densities for longer work zones setup. This aligns with the common practice of implementing late merge control for work zone works scheduled during peak periods. Moreover, mainline merge metering yielded more promising results in terms of density. It should be noted that impacts on surface network were not evaluated when applying temporary ramp metering. Accordingly, implementation of temporary ramp metering strategies should be considered with full account for impacts over freeway and surface street networks. As for off-peak traffic conditions, late merge control and temporary ramp metering did not impact density. On the other hand mainline merge metering and early merge control yielded higher densities with mainline merge metering having the worse impact during off-peak conditions. In addition, minimal impact was found for work zone length over density, which indicates, from an operational perspective, that the merge control strategy is more important than work zone length.

Lane changes total. The total number of lane changes that have been completed during the simulation period on the links occupied by the work zone, in addition to links comprising a stretch of the interstate upstream and downstream of the work zone with the same length was considered. A lane changes total index (ILC,Link) was calculated to represent the lane changes in comparison to those of baseline conditions. Figure 8 illustrates the ILC,Link by work zone length for both peak and off-peak traffic. A positive ILC,Link indicates excessive lane changing, while a negative value indicates less lane changes compared to baseline conditions.

It is observed that lane changes tend to decrease during peak traffic and as the work zone gets longer. However, during off-peak traffic, lane changes tend to be unchanged except under the mainline metering strategy. In addition, during peak traffic, mainline metering strategy yields the smallest number of lane changes; however, during off-peak traffic, it yields the highest number of lane changes.

It should be noted that excessive lane changing is a key traffic safety hazard. Accordingly, simulation results indicate that under peak traffic conditions, early merge control yields the least hazardous merge control strategy, while late merge control strategy is the most favorable under off-peak traffic conditions. It is also observed that all strategies have consistent impact up-to work zone length of 2000 ft (610 m); however, when work zone length exceeds that limit, significant improvements are observed, indicating that longer work zones yield fewer lane changes.

Speed average (Km/Hour). The link-based Speed Average MOE represents the average speed of vehicles that have completely traversed the link. This MOE was collected for the links occupied by the work zone, in addition to links comprising a stretch of the interstate upstream and downstream of the work zone with the same length. A link speed average index (IS,Link) was calculated to compare the changes in link speed average compared to baseline conditions. Figure 9 illustrates the plot of this IS,Link by work zone length for peak and off-peak traffic. A positive value indicates improved speeds, while a negative value indicates slower traffic.

As showed in Figure 9, all strategies yield better performance under peak traffic compared with off-peak traffic conditions. In addition, temporary ramp metering improved the speed average when implemented, due to its inherent feature of providing priority to mainline vehicles in the expense of those who want to get on the freeway.

Speed average is a key mobility indicator on interstate links, and under such terms temporary ramp metering and mainline metering strategies have the most favorable mobility impact under peak traffic conditions, and early merge control has the most adverse mobility impact under such conditions. However, under off-peak traffic conditions, mainline metering yields the most unfavorable mobility impacts, while all other strategies have minimal impact on mobility. Moreover, it is observed that work zone length has no or minimal impact on speed average under mainline metering strategy. In addition, under temporary ramp metering, longer work zones yield better speed averages. As for late and early merge control, the study results show that longer work zones decreases speed average.

Delay travel per vehicle (Seconds/Vehicle). The link-based MOE delay travel per vehicle represents the average time that vehicles were delayed on the link. A link delay travel per vehicle index (IT,Link) was calculated for the links occupied by the work zone, in addition to links comprising a stretch of the interstate upstream and downstream of the work zone with the same length, to indicate the change in delay travel per vehicle in comparison to baseline conditions. Figure 10 illustrates the plot of IT,Link by work zone length for peak and off-peak traffic.

Results aligned with that of link-based speed average reported earlier. More specifically it is observed that all strategies yield better performance under peak traffic compared with off-peak traffic conditions, except for mainline metering where performance worsened under off-peak traffic conditions. In addition, temporary ramp metering minimized delay travel when implemented by providing priority to mainline vehicles in the expense of vehicles merging from the ramp. Delay travel per vehicle complements the speed average MOE as key mobility indicators on interstate links, and under such terms it is observed that temporary ramp metering and late merge control strategies have the least impact on delay travel under peak traffic conditions, and early merge control has the most adverse mobility impact under such conditions. However, under off-peak traffic conditions, mainline metering yields the most unfavorable delays, while all early merge control have the least impact on delay travel. In addition, it is observed that work zone length has no major impact on delay travel.

Emissions total CO (Grams/Km). The link-based MOE emissions total CO, in grams/Km, represents the total CO emissions per mile produced by all vehicles on the link. This MOE is the key environmental impacts indicator, and was collected for the links occupied by the work zone, in addition to links comprising a stretch of the interstate upstream and downstream of the work zone with the same length. An emissions total CO index (IE,Link) was calculated to indicate the change of emissions in comparison to baseline conditions. Figure 11 illustrates a plot of IE,Link by work zone length for peak and off-peak traffic. A positive surge factor indicates worse emissions, and a negative value indicates improvement.

For all strategies except mainline metering, tested merge control strategies resulted in a change of ±10% of preconstruction values. However, mainline metering resulted in an increase of 70% - 80% compared to preconstruction conditions. As expected, all strategies yield better environmental performance under off-peak traffic except for late and early merge controls due to decreased speeds and hence lesser engine loads. In addition, early merge control minimized emissions total CO when implemented both under peak and off-peak traffic conditions. In addition, it is observed that longer work zones yield better emissions total CO, which indicates that the longer the work zone the more favorable the environmental impacts.

Network-Based MOEs

Network-based MOEs provide a global understanding of the tested merge control strategies at interstate work zones. Two MOEs were selected to describe the mobility impacts of tested merge control strategies, namely speed average and delay travel total. The following sub-sections discuss these two MOEs and relevant observations. Surge Factors were used once again to indicate the change for each network MOE with respect to implemented strategy by work zone length.

Speed Average (Km/Hour). The network-based Speed Average MOE represents the average speed of vehicles traveling in the network. A network speed average index (IS,Net) was calculated to illustrate the percentage change of all vehicles speed average from baseline conditions. Figure 12 illustrates the plot of this IS,Net by work zone length for peak and off-peak traffic conditions. As mentioned earlier, a positive surge factor indicates lower network speed average and a negative value indicates faster network traffic.

It was observed that all strategies yield better performance under off-peak traffic compared to peak traffic conditions. In addition, under peak traffic conditions, mainline metering had the least impact on the network average speed, while temporary ramp metering had the worst impact. Vehicles speed was reduced by 25% - 45% (36 - 27 mph respectively) when merge control strategies were implemented on a work zone. As for off-peak traffic conditions, results indicate that nearly all strategies had no to minimal impact on the network speed average. It is also observed that work zone length had minimal to no impact on the network speed average.

Delay travel total (Hours). The network-based delay travel total MOE represents the total time that all vehicles were delayed in the network. A network delay travel total index (IT,Net) was calculated to illustrate the percentage change of delay travel total from baseline conditions. Figure 13 illustrates a plot of IT,Net by work zone length for peak and off-peak traffic.

As expected, all strategies yield better performance under off-peak traffic compared to peak traffic conditions. In addition, under peak traffic conditions, results were consistent with the network speed average as mainline metering had the least impact on the delay travel total, while temporary ramp metering had the worst impact. Delay travel total surged by 100% - 300% (equivalent to 28 - 55 seconds/miles traveled, respectively) when merge control strategies were implemented on a work zone during peak periods. Same behavior was observed under off-peak traffic conditions with a surge ranging between 50% - 225% (equivalent to 20 - 45 seconds/miles traveled, respectively). It was also observed that work zone length had minimal to no impact on the delay travel total.

CONCLUSIONS

Four merge control strategies for interstate work zones of various lengths were tested in the CORSIM microsimulation platform, namely late merge control, early merge control, mainline merge metering, and temporary ramp metering. The simulation model accounted for the MUTCD requirements as well as driver/vehicle behavior in work zones and considered peak- and off-peak traffic conditions. Five link-based MOEs and two network-based MOEs were analyzed and reported.

Results indicated that under peak traffic conditions the mainline merge metering strategy is expected to yield the most favorable safety impact represented by fewer lane changes, favorable mobility impacts represented by lesser speed reductions, and improved LOS represented by lower densities. However, mainline merge metering would result in undesirable impacts on the environment, represented by higher emissions total CO. Under off- peak traffic conditions, late merge control surpassed all other strategies and yielded the most favorable impacts, except for density, where temporary ramp metering resulted in similar impact on LOS. Moreover, results indicate that under peak traffic conditions several strategies would yield acceptable impacts. Hence, this reveals that nighttime scheduling of interstate maintenance works may not be the most favorable option.

Overall, this study provides practitioners and researchers a comprehensive and comparative evaluation of the four most utilized merge control strategies, under both peak and off-peak traffic conditions within a work zone setting. In addition, the study provides comprehensive information on key parameters to calibrate and setup CORSIM in order to model work zone conditions at freeway facilities. One of the contributions of this work is to explore and test creative ways to model the strategies within the constraints of the CORSIM simulation platform.

Given the findings of this study, it is highly recommended to reexamine scheduling practices of interstate work zones in a future study. Such a study should consider measures for LOS, user satisfaction, and environmental impacts resulting from implemented merge control strategy and choice of scheduling periods. In addition, additional work is recommended to evaluate merge control strategies for work zones present under different configurations such as two-lane freeways and arterial settings.

REFERENCES

1. Wei, H., Pavithran, M., Yi, P., Yang, Q. and Zeng, Q.-A. (2010) Synthesis of Unconventional Dynamic Merge Metering Traffic Control for Work Zones. Open Transportation Journal, 4, 52-60. http://dx.doi.org/10.2174/1874447801004010052
2. Chitturi, M.V., Benekohal, R.F. and Kaja-Mohideen, A.-Z. (2008) Methodology for Computing Delay and User Costs in Work Zones. Transportation Research Record, 2055, 31-38. http://dx.doi.org/10.3141/2055-04
3. Tang, Y. and Chien, S.I.-J. (2008) Scheduling Work Zones for Highway Maintenance Projects: Considering a Discrete Time-Cost Relation. Transportation Research Record, 2055, 21-30. http://dx.doi.org/10.3141/2055-03
4. McCoy, P.T. and Pesti, G. (2001) Dynamic Late Merge-Control Concept for Work Zones on Rural Interstate Highways. Transportation Research Record, 1745, 20-26. http://dx.doi.org/10.3141/1745-03
5. Federal Highway Administration (FHWA) (2012) Manual on Uniform Traffic Control Devices (MUTCD). 2009 Edition, FHWA, Washington DC.
6. Ramadan, O.E. and Sisiopiku, V.P. (2015) Bottleneck Merge Control Strategies for Work Zones: Available Options and Current Practices. Open Journal of Civil Engineering, 5, 428-436. http://dx.doi.org/10.4236/ojce.2015.54043
7. Hardy, M. and Wunderlich, K.E. (2009) Traffic Analysis Tools Volume IX: Work Zone Modeling and Simulation—A Guide for Analysts. Report № FHWA-HOP-09-001. Federal Highway Administration, Washington DC.
8. Alabama Department of Transportation (ALDOT). Alabama Traffic Data, 2014. http://algis.dot.state.al.us/atd/
9. Dowling, R., Skabardonis, A. and Alexiadis, V. (2004) Traffic Analysis Toolbox Volume III: Guidelines for Applying Traffic Microsimulation Software. Report № FHWA-HRT-04-040. Federal Highway Administration, Washington DC.
10. Holm, P., Tomich, D., Sloboden, J. and Lowrance, C. (2007) Traffic Analysis Toolbox Volume IV: Guidelines for Applying CORSIM Microsimulation Modeling Software. Report № FHWA-HOP-07-079. Federal Highway Administration, Washington DC. [Citation

Time(s):4]
11. Schroeder, B.J., Cunningham, C.M., Findley, D.J., Hummer, J.E. and Foyle, R.S. (2010) Manual of Transportation Engineering Studies. 2nd Edition, Institute of Transportation Engineers, Washington DC.
12. Shoaib, L. (2002) Calibration of FRESIM for Work Zones Using Field Observable Parameters. Master of Science Thesis, University of Cincinnati, Cincinnati.
13. Vadakpat, G., Stoffels, S. and Dixon, K. (1999) Calibration and Validation of CORSIM for Work Zones. ITE Journal, 69, 12-14.
14. American Traffic Safety Services Association (ATSSA) (2008) Maintenance Work Zone Safety: Pocket Guide of MUTCD Guidance on Temporary Traffic Control. ATSSA, Fredericksburg.
15. Zegeer, J.D., Vandehey, M., Blogg, M., Nguyen, K. and Ereti, M. (2008) Default Values for Highway Capacity and Level of Service Analyses. Transportation Research Board, Washington DC. http://worldcat.org/isbn/9780309099318 http://dx.doi.org/10.3141/2071-05
16. Federal Highway Administration (FHWA) (2015) HPMS Field Manual—Appendix N: Procedures for Estimating Highway Capacity, 2011. http://www.fhwa.dot.gov/ohim/hpmsmanl/appn.cfm
17. McTrans Center (2011) CORSIM Users Guide. University of Florida, Gainesville.
18. Transportation Research Board (TRB) (2010) Highway Capacity Manual 2010. TRB, Washington DC.

CHAPTER 3

Modulating Traffic Signal Phases to Realize Real-Time Traffic Control System

Rajendra S. Parmar[1], **Bhushan H. Trivedi**[2]

[1]Gujarat Technology University (GTU), Ahmedabad, India

[2]GLS Institute of Computer Technology, Ahmedabad, India

ABSTRACT

This paper proposes innovations to address challenges emanating from road traffic congestion. Improving economies create more car owners resulting in increased automobile manufacturing, increased vehicle population giving rise to higher emission of CO_2 resulting in traffic congestion. Congested traffic has idling vehicles which emit higher CO_2 and pollution. Besides,

Citation: Parmar, R. and Trivedi, B. (2017), "Modulating Traffic Signal Phases to Realize Real-Time Traffic Control System". *Journal of Transportation Technologies*, 7, 26-35. doi: 10.4236/jtts.2017.71002.

traffic congestion increases turnaround time, delivery time, commuting time and related logistical aspects. Commuting time negatively impacts working hours. Unless the traffic congestion is mitigated, the economy will take a beating creating a vicious ecology cycle. Building new roads, bridges or reconditioning of infrastructure is not always the best possible solutions. Efficient traffic management is a key to country's economic growth. Various analytical models are employed to study, appreciate traffic congestion. The paper studies these models to infer that real time approach is the only solution. Several approaches are being worked on and few commercial systems too are available. These systems provide traffic information for course correction. However, it has latency and hence deviates from real time environment. Traffic congestion being highly dynamic in nature, it necessitates real time solution with real time inputs. It is proposed to integrate Real time traffic data with the traffic signal thus modulating the cycle timings at every junction. Deviation from static asymmetric cycle timing is implemented by assigning green phases based on density of vehicles. With minimalistic infrastructure and negligible incremental cost, the paper not only proposes to address traffic congestion but also paves the way for capturing traffic offenses, vehicle tracking and toll collection. The research is imminently realizable and makes a strong case for a PPP (Public Private Partnership) project.

Keywords: Vehicular Traffic, Vehicular Congestion, Detection System, Vehicular Congestion Detection System, Vehicle Mobility Sensors, Traffic Signals, Intelligent Traffic Signals, Dynamic Traffic Assignment

INTRODUCTION

The location accuracy from Global Positioning System (GPS) device is continuously improving; from 15 meters we have accuracy of 3 meters on systems that deploy WAAS (Wide Area Augmentation System). GPS uses at least three satellite data and communicates with GPS enabled receivers. WAAS uses a network of ground-based reference stations. Measurements from ground based stations along with GPS are sent to the geostationary WAAS satellites. WAAS satellites then sent the corrected data which is received by the WAAS enabled GPS devices.

PNDs (Personal Navigation Devices) are best suited for traffic application. The advantages are large displays, keyboards, arms length viewing and usability while driving. Besides, PND have room to accommodate speed, upcoming street, time of arrival and other parameters. However, in spite of these factors smart phones are preferred over PND. The mapping assistance is available through mapping giants like TeleNav, NavTeq (Navigation Technologies), Google, AT & T. The PND based navigation systems is available at USD $70 with a yearly update cost of USD $70 approx whereas the smart phone besides the device cost & data connectivity charges is available for USD $50 for the navigation app. Analog traffic receivers have an update frequency of 300 seconds. Garmin live traffic updates 1000 messages every 2 minutes on smart phones. Garmin devices with High Definition (HD) digital offer navigation services with live traffic updates for approx. more than USD $300. The traffic update frequency is 30 seconds.

Motion X GS drive offers community contributed road alerts and reporting including accidents, hazards, police location, weather, and more. Tom-Tom pro- vides traffic information by integrating variety of traditional data sources like governmental or third party data through road sensors or through mobile phone users. The system merges this information, analyses it and sends the data to users every three minutes.

In short, data come from millions of Garmin device owners, millions of cellular phone owners, incident reports, live information through radio feeds, news stations, NAVTEQ Traffic Supply and historical traffic data from each of these agencies. Some of the systems require downloading with periodic refreshing and work without the need for continuous connectivity.

RELATED WORK

Proposed efficient Intelligent Traffic Control System (ITCS) system [1] . The digital-logic based principle being "a car can only move ahead if there is space for it" and "the signal remains green until the present cars have passed". Sensors are placed entry and exit of a road segment or a road leading to a traffic junction. These sensors provide information about the vehicles entering the junction and leaving the junction. This information is sufficient to know the number of vehicles on the road segment. Based on the vehicle population, the traffic signal phases are computed. The challenge here is fixed sensors. Fixed sensors have higher capital costs, installation cost and maintenance cost. It requires 24 × 7 power supply. Broadly speaking is prohibitively costly as all the traffic signals have to be equipped with the sensors.

Defines guidelines for Anthropocentric design for self-driving vehicles [2] . Presented intelligent Traffic Light Controllers (ITLC) based on microcontroller and microprocessor [3] . It has communication interface through which cycle times can be changed dynamically. Traffic signal related performance criterions can be logged and stored on a central server.

Google traffic displays traffic in real time by acquiring data from its users and calculating the speed [4] . On the map, Google shows green if traffic is moving along; yellow, if some traffic; red, if appreciable congestion.

GNSS-1 is the first generation system, which integrates existing GPS with other technologies. In USA, the satellite based component is the Wide Area Augmentation System (WAAS), in Europe it is the European Geostationary Navigation Overlay Service (EGNOS), and in Japan it is the Multi-Functional Satellite Augmentation System (MSAS). GNSS-2 is the second generation of system. These systems collate vehicle velocities. Commercial system like Tom Tom & NavTeq capture PVT (Position, Velocity & Time) to provide traffic information.

Academicians, researchers and corporations have been working on traffic control systems with a view to optimize the traffic signal phases, reduce signal idle time and address traffic congestion. The technology is based on ATCS (Ada- ptive Traffic Control System). The systems are evolving and continuously improvising to mitigate the traffic congestion problems. Before considering Adaptive Traffic Control Systems (ATCS) deployment, it is important to understand the reasons an ATCS is required.

Contemporary Systems

The software is based on Windows and OS and all of them have application software of their own. The traffic controllers used are Siemens M 50 series or 2070, Peek 3000E, Model 170, SITRAFFIC C8xx, C9xx, Econolite ASC/2 and NEMA AWA Delta 3N. The installation cost of ATCS per junction is approx. USD $65,000.

SCOOT (Split Cycle Offset Optimization Technique) is a real time application for controlling traffic signals based on ATCS (Adaptive Traffic Control System) [5] . All traffic signals are connected and coordinated to allow smooth traffic flow. It deploys infrastructure information with the in-road sensors to compute traffic flow and traffic density. It offers all the information required to device vehicle routing thereby minimizing vehicle delays.

SCAT (Sydney Coordinated Adaptive Traffic System) is an ATCS using complex algorithms to generate traffic signal phases so as to minimize delays [6] . SCAT is a real time solution adjusting signal phases based on traffic area rather than vehicle count. The data is gathered through in-road loop sensors, which provides computing traffic density. SCAT in turn determines the duration of signal phases. SCAT provides operations and maintenance monitoring including; real time alarm monitoring, space and time diagrams as well as support from historical data which can help future planning.

Urban Traffic Optimization by Integrated Automation (UTOPIA) is integrated with SPOT to account to changes at the network level (Mauro and DiTaranto 1990). The parameters considered are vehicle detection, vehicle flow rate, traffic intensity, speed, road occupancy, density and type of vehicle. FHWA (Federal High-Way Administration), USA, developed ACS Lite which is widely deployable, at much reduced cost than traditional installation, operations and maintenance [7] . The system has CORSIM simulator and employs signal controllers from Eagle, Econolite, McCain and Peek controllers. The installations began around 2005. RHODES is a ATCS which began its implementations around 1990 [8] . RHODES predicts traffic levels taking sensors inputs from inductive loop sensors, RADAR, SONAR or video inputs. It then converts into lanes data and computes signal phases and communicates with server. RHODES also computes dynamic shortest real time to divert traffic from congested area to lean traffic area.

The comparison of the two systems with largest market share is tabulated in Table 1.

ANALYTICAL MODELS

Dynamic traffic assignment (DTA) is a subject of interest for the last decade and is yet maturing. The traffic networks are probabilistic and uncertain, various models were proposed. Broadly these models, model travelers' decision making, probabilistic travel time and random perception errors of individual travelers. Lighthill, Whitham, Richard introduced the continuum model (LWR model) based on fluid dynamics [9] . It proposes a function between traffic density and speed to capture the characteristics of traffic congestion formation. A congestion factor is evolved based on traffic data and principles of traffic congestion formation. The first-order continuum flow model, on dense traffic with equilibrium where k and q are density and flow respectively which depend on time t and position x

$$\frac{dk(x,t)}{dt} + \frac{dq(x,t)}{dx} = 0$$

The model suggests preferred speed for vehicles. It is flawed, because, when passing is allowed, the preferred speed for each vehicle varies over time. Also the desired speeds among a group of vehicles vary.

PW Model: Payne-Whitham higher model introduces additional terms in the equation of motion proposed by LRW model, Navier-Stokes equation of motion for one-dimensional compressible flow with pressure and a relaxation term [9] [10] :

Table 1.Comparison of contemporary Sysems.

Parameters	SCOOT	SCAT	Rhodes
Installations	14 Countries; 200 systems	27 Countries; 37,000 Intersections	
Developer	1) IMTECH Traffic & Infra UK Ltd 2) Siemens Traffic Controls 3) TRL Limited (Transport Research Laboratory) funded by DfT and the SCOOT suppliers	1) Australia 2) New South Wales Government, Australia	Research team at University of Arizona
Year	Early Eighties; 1980	1975	
Description	Split Cycle Offset Optimization Technique	Sydney Coordinated Adaptive Traffic System	Real-time Hierarchical Optimizing Distributed Effective System
Type	On Line	On Line	Off Line
Software	1) SCOOT Kernel, UTC software 2) Microsoft Windows	1) Client Server Technology 2) Windows OS	Algorithm can be implemented on any platform
Performance	Reduces delay by 20%	Reduces delay by 5% to 45%	
Detectors Technologies	On-Street Detectors; Inductive loops	On-Street Detectors; Inductive loops	Inductive detectors, Video, SONAR, RADAR
Controllers	SIEMENS	1) SCAT compatible traffic controller 2) 2070 and 170 traffic signal controllers	
Deployment Cost	Very High	Very High	Very High
Shortest Path	No	No	Yes
Comprehensive Traffic Information	ASTRID (Automatic SCOOT Traffic Information Database)	Historical Data	No
Fault Monitoring & Management	Fault detection logic	Real Time Alarm System	Not available
Features	1) Bus priority 2) Manages pedestrians at junctions 3) Estimate fuel emission 4) INGRID (Integrated Incident Detection)	1) Control air pollution 2) Fuel consumption, delays and reliability 3) Real time alarm monitoring, space and time diagrams	1) Real Time Dynamic shortest path

$$\frac{dv}{dt} = -v \cdot \frac{dv}{dx} - \lambda\left(v - Ve\left(K\right)\right) - \frac{1}{k} \cdot \frac{dP}{dk} \cdot \frac{dk}{dx}$$

where v is traffic speed, Ve(k) is the equilibrium speed-density relationship, P(k) is traffic pressure, and λ is a coefficient.

Kalman filtering approach implements linear flow model [11] . Though here too the observed density computed has inherent inaccuracies. Phillips' Model incorporates mass conservation, momentum conservation, and energy conservation. Kuhnes' Model Incorporated sound speed and viscosity.

Kerner and Konhauser's model proposes that Initial homogeneous traffic flow can spontaneously result in regions of high density and low average speed. These high-density regions can move with or against the flow of traffic. Michalopoulos model does not require the use of an equilibrium speed-density relationship. Zhang proposed a non-equilibrium traffic flow model which is based on both empirical evidence of traffic flow behavior and basic assumptions on drivers' reaction. Treiber Derived that in free flow, traffic information propagates downstream, while in congestion it travels upstream. Even though this method takes into account the way in which information propagates in traffic streams, it does not make use of any flow model [10] [11] .

Short Summary

The study helps understand that (Table 2):

- Traffic congestion is spatial-temporal conglomeration of traffic volume in finite time and space,
- Traffic is multi-vehicle system with infinite equilibrium states (Not a unique equilibrium state),
- Vehicular traffic flow is nonlinear dynamic process,
- High-order models involve more complex partial differential equations, more variables which increases computational complexity and are more difficult to calibrate and implement,
- It is believed, high-order models, do not improve LWR model,
- Research till date has not been able to identify principles of traffic congestion OR conclusive conditions responsible for forming traffic congestion,
- Under the circumstances, congestion status of traffic flow can be evaluated in real time only [9] .

PROPOSITION

Traffic signal works on predetermined cycle times with each direction having green orange and red phases. These cycle times can be programmed to change or can be manually changed to cater to at peak and lean periods. However, a real time, dynamic modulation of traffic cycle time is desired to maximize the traffic flow. Has demonstrated need to utilize unused green phases [12] . It is proposed to assigning green phases proportional to the vehicular density in the respective direction. The strategy translates into uniform dispersion of vehicles rather than uniform cycle time to each direction. This will result in decongesting congested vehicles and evenly distributing vehicular densities.

Table 2. Summary of related research work.

Model	Limitations
LRW (Lighthill, Whitham, Richard) OR Continuum Model	• Model suggests Preferred speeds • Different vehicles have different preferred speeds • Predicts abrupt speed change when a vehicle passes through a shockwave, an action that is unrealistic in real world. • LWR model fails to predict instabilities of stop-start traffic • LWR model assumes zero reaction time which does not happen in real world
PW Model: Payne-Whitham	• Like LRW model, It does not remove all the shock waves • Vehicles are expected to adjust their speeds in relation to the rear vehicles, while in reality vehicles respond to thread • Proposed wave speeds are greater than traffic speeds. Again primarily because it derives the recommendations from rear vehicles
Kalman filtering approach	• Velocity of an individual vehicle is assumed to correspond to the velocity at that given location • The velocity-to-density conversion, introduces error • Measurement error in the velocity

The important parameters to compute are listed in Table 3.

The paper integrates vehicle registration, mapping application, traffic monitoring and modulating the traffic signals (Table 3).

Vehicle Mapping

Vehicle mapping assigns a device ID to a vehicle. Extreme care is taken to maintain users' privacy and data confidentiality. Besides, vehicle manufacturer, model, vehicle area (length, breadth), fuel type, fuel economy and other parameters of importance are logged. The make and model, fuel type, distance travelled are parameters for computing and monitoring pollution data from the vehicular emission. The time stamped data with latitude & longitude, road name builds a history, behavior and knowledge database about the vehicle, road segment. The ID also identifies vehicles as emergency vehicle, transport vehicle, commercial, public vehicle or private vehicle.

Mapping Application

Mapping application is intended to capture infrastructure characteristics like length of road segment, number of lanes that includes width of road segment, right of travel direction-unidirectional or bidirectional and quality of road segment to decide average speed of vehicle on the road segment. The application computes how many vehicles can be present in the road segment. Besides it logs the latitude and longitude of the ends of a road segment, name of the road segment which enables the graphic application to render the map. A survey vehicle is equipped with the infrastructure information capturing application drives through the entire city with a data entry team. The application prompts for nomenclature (Location description) every time there is a change in lat/long. The average speed is captured from the length of the road segment and time taken to traverse the same. The speed is related to the quality of road segment. The user enters number of lanes, right of direction. The user also enters the number of converging roads and traffic signals at both ends of the road segment.

Table 3. Traffic junction parameters of importance.

Parameter	Value	Unit
Min. Green Phase Time	t_{min}	s
Max. Green phase time	t_{max}	s
Time to cross junction	t_t	s
Vehicles at junction	v1	No
Capacity to hold Vehicles	c1	No
Vehicles at next junction	v2	No
Capacity to hold Vehicles	c2	No

Traffic Monitoring

The application periodically captures lat/long. The periodicity is limited to remove data redundancy and only create useful data, rendering data compression. The vehicle position is updated in the server. The speed computation is done by taking two reasonable samples of positions with a time stamp. The sample provides distance and time to travel the distance, thus giving average speed. Vehicle density is computed by adding up all the vehicles whose position falls between the lat/long of two junctions. Infrastructure details are obtained from the database detailing allowable speeds on the road segment, length of the road segment and the capacity of the road segment besides other parameters. Congestion index is computed by comparing the present vehicular density to the capacity of the road segment. Congestion index of 0.8 to 1.0 means heavy congestion, if the index is between 0.6 to 0.8 it would mean moderate traffic and below 0.6 is free flowing traffic. The server responds with all the relevant information to assist the driver.

Modulating the Traffic Signals

Traffic is controlled intelligently after accurate, real time data is obtained. So far all the directions were assigned fair, equal signal phases. However, many a times when there is no contest or contention for a direction because there are no vehicles. Hence asymmetric signal phase is proposed. The proposal is to give priority to a direction with maximum number of waiting vehicles. However, there is a possibility of starvation amongst directions with fewer vehicles. Hence waiting time is incorporated along with number of vehicles. Waiting time is summation of waiting time of all vehicles at a junction. The intension is to normalize waiting time of vehicles. This will ensure a truly fair assignment of signaling to different directions. The central server has data of all the vehicles on the road which includes the road segment they are in, average speeds and vehicles waiting at a traffic junction besides other relevant information.

1). The algorithm is designed to service the direction with largest number of vehicles waiting for the longest time expressed as:

$$\tau = \sum_{1}^{N} \left(\int_{T1}^{T2} dt \right)$$

τ = Cumulative waiting time (CWT) of vehicles for a direction; T1 = Entry time of vehicle; T2 = Time at computation; N = Number of vehicles.

The algorithm continuously evaluates green phase and assigns to a direction. While doing so, enough care is taken not to starve the other directions indefinitely. The upper and lower bound of green & red phases ensures that the direction with dense vehicle population is assigned a red phase and the sparsely dense direction too gets its share of green phase.

2). The green phase is assigned only when the road segment ahead has capacity to accommodate additional vehicles OR the green phase timing is defined based on how many vehicles it can accommodate.

3). There might be a case when all the signal phases are red indicating heavy congestion. It allows the road ahead to create enough buffer to accommodate additional vehicles. This ensures the system allows the congestion to build up to a manageable level and not frustrating level of congestion.

FUTURE WORK PROPOSED

[13] demonstrated that violations and offenses like speeding, license plate cloning, vehicular robbery, driving with influence of alcohol can be detected. The proposed system in this paper detects over speeding vehicles. The device installed on the vehicle, has a unique ID. Addressable vehicles are monitored for violations like over speeding, jumping traffic signals, wrong parking or long parking hours. An over-speeding vehicle is detected immediately and so is the detection of a breakdown. Needless to say, the toll collection is automated. The system collets toll based on travel zones. With exact vehicle location, vehicle speed, speed limit of the road segment and other parameters being available, post accident analyses becomes a trivial matter. Besides the routine display parameters, the dash board can indicate distances of neighboring vehicles, their speeds, cross roads ahead with a notification of green, red or orange situation. Lastly, real time traffic control system can also pave the way to eliminate signaling systems at every traffic junction by implementing signaling indications in each vehicle dash board directly. This will make the traffic signaling systems redundant, resulting in large savings in its capital as well as operational & recurring maintenance costs.

REFERENCES

1. Bilal, J.M. and Jacob, D. (2007) Intelligent Traffic Control System. IEEE International Conference on Signal Processing and Communications (ICSPC 2007), Dubai, 24-27 November 2007.
2. Mladenovic, M.N., Abbas, M. and McPherson, T. (2014) Development of Socially Sustainable Traffic-Control Principles for Self-Driving Vehicles: The Ethics of Anthropocentric Design. IEEE International Symposium on Ethics in Science, Technology and Engineering.
3. Chavan, S.S., Deshpande, R.S. and Rana, J.G. (2009) Design of Intelligent Traffic Light Controller Using Embedded System. Second International Conference on Emerging Trends in Engineering and Technology, ICETET.
4. http://www.theconnectivist.com/2013/07/how-google-tracks-traffic/
5. http://www.scoot-utc.com/
6. http://www.scats.com.au/
7. Shelby, S.G., Bullock, D.M., Gettman, D., Ghaman, R.S., Sabra, Z.A. and Soyke, N. (2008) An Overview and Performance Evaluation of ACS Lite—A Low Cost Adaptive Signal Control System. 87th TRB Annual Meeting, Washington DC, January 2008.
8. RHODES—Real Time Hierarchical Optimized Distributed Effective System. http://ocw.nctu.edu.tw/course/sc011/2012-08-23-1.pdf
9. Zhang, W., Tan, G.Z., Ding, N. and Wang, G.Y. (2012) Traffic Congestion Evaluation and Signal Control Optimization Based on Wireless Sensor Networks: Model and Algorithms. School of Computer Science and Technology, Dalian University of Technology, Dalian.
10. http://people.umass.edu/ndh/TFT/Ch15%20High.pdf
11. Herrera, J.C. and Bayen, A.M. Departamento de Ingeniería de Transporte y Logí- stica, Pontificia Universidad Católica de Chile, Chile, Systems Engineering, Depart- ment of Civil and Environmental Engineering, University of California, Berkeley, United States; Incorporation of Lagrangian Measurements in Freeway Traffic State Estimation.
12. Parmar, R.S. and Trivedi, B. (2014) Real Time Computation of Optimal Signal Timing to Maximize Vehicular Throughput for a Traffic Junction. 3rd International Conference on Eco-Friendly Computing and Communication Systems (ICECCS), ICECCS 2014, Mangalore, 18-21 December 2014.

13. Ou, G.Y., Gao, Y. and Liu, Y. (2012) Real-Time Vehicular Traffic Violation Detection in Traffic Monitoring Stream. IEEE/WIC/ACM International Conferences on Web Intelligence and Intelligent Agent Technology.

CHAPTER 4

Traffic Regulation at Critical Intersections: A Case Study of Odole Intersection, Akure, Ondo State, Nigeria

Olumuyiwa Samson Aderinola, David Oladotun Owolabi

Department of Civil and Environmental Engineering, Federal University of Technology, Akure, Nigeria

ABSTRACT

Traffic congestion on major roads consequent upon existing bottlenecks at intersections is a major problem in Akure Metropolis. To change this trend, this research was carried out in order to design effective traffic control mechanisms at the "troubled spots" in the metropolis. Odole intersection was identified as one of the critical intersections during a reconnaissance

Citation: Aderinola, O. and Owolabi, D. (2016), "Traffic Regulation at Critical Intersections: A Case Study of Odole Intersection, Akure, Ondo State, Nigeria". *Open Journal of Civil Engineering*, **6**, 94-104. doi: 10.4236/ojce.2016.62009.

survey and as such, selected for study. Data on geometric features were collected using Oedometer and Google Earth software. Peak and off-peak traffic volume data were collected between 7:30 and 8:30 am and between 12.00 noon and 1.00 pm respectively every other day using Cine camera. Furthermore, discharge headway and delay data were collected using stop watch. The geometric and traffic data collected were analysed using Microsoft Excel. An appraisal of Odole Intersection indicated that the major contributors to traffic are Motorcycles (70.88%) and Passenger Cars (28.72%). Other modes of transportation account for about 0.4% of vehicles traversing the intersection. The critical traffic volume at the intersection was over 4000 veh/hr and the average delay was 22 seconds. An Average delay of 22 seconds at the intersection was an indication that the operating level of service was C (i.e. fairly stable traffic condition with average delays attributable to traffic control by personnel). By juxtaposing the results of geometric and traffic data analyses with the pros and cons of various traffic control mechanisms, traffic control by signalisation was selected and designed to suit Odole intersection. In tandem with the results of this research, appropriate measures have been recommended to ameliorate traffic and commuting problems in the metropolis.

Keywords: Traffic Congestion, Traffic Data, Signalisation

INTRODUCTION

The history of man is tied to history of movement. Transportation is essential to social progress; transportation itself is a social progress because it has been, throughout history, the way in which not only goods and services are moved around but ideas are exchanged among people [1] . However, the roses of transportation might grind to a halt as a result of bottlenecks occurring at intersections. An intersection is a general area where two or more roads cross at the same or different elevations [2] . Interruption of traffic flow is a daily experience in urban centers of the world and it occurs primarily at intersections.

Road traffic jams continue to be a major problem in most cities around the world, especially in developing regions resulting in increased fuel wastage and monetary losses [3] . The forecast of Global Traffic Volume (GTV) shows that road traffic jams would quadruple between 1990 and 2050 [4] . This type of traffic growth pattern, as envisaged by the end of year 2050, is an indication of what the future congestions portends for people living in urban environment. Effective traffic control mechanisms are therefore pertinent to curb the menace. Traffic control mechanisms at intersections include prioritization, traffic personnel control, channelization, signalization, rotary intersections and by grade separation (interchanges). Each of the aforementioned control mechanisms has its areas of applicability and limitations.

When traffic flows at an intersection are relatively low, adequate control can be achieved using the priority option. As flow levels increase, with this intersection type, delays/queue lengths become excessive and some alternative form is required. While grade-separated junctions may be the preferred option at high flow levels, the expense involved may be prohibitive. For this reason, particularly in an urban setting, at-grade roundabouts or signalised intersections become viable junction options at levels of flow above those suitable for priority control [5] . If the cost of land is an important factor, traffic signals will be preferred since land requirements for a standard 3 or 4-arm conventional roundabout would be greater [6] . However, grade separation is justified when a full control of access is to be achieved on roads with intersections; where volumes to be catered for would require the design of an intersection at grade of unreasonable size; in accident prone regions. Grade separation could also be justified in situations whereby topography hinders the feasibility of other forms of control and when huge economic losses could be incurred due to intersectional frictions at grade [7] [8] .

DESCRIPTION OF STUDY AREA

Akure metropolis, used as the study site is the capital city of Ondo State and it is located in south western Nigeria (as shown in Figure 1). Ondo state was created in 1976 and shares boundaries with Ekiti and Kogi States in the north, Edo State in the east, Delta State and Atlantic Ocean in the South. Akure is located in the northern part of the state on latitude 7°15'0"N and longitude 5°11'42" and has an area of 30.02 square kilometres [9] .The city has a population of 340,021. The population growth rate in the metropolis is about 3.5% per annum [10] .

Climatic Characteristics of Akure Metropolis

The climate is hot and humid with two distinct seasons, the rainy and dry seasons. The rainy season lasts for seven months (April - October) and the dry season for five months (November - March). The rainfall is about 1524 mm per year and the atmospheric temperature ranges between 28°C and 31°C with a relative humidity of 80 percent [11] . Akure is located in the forest zone with abundant forest resources and it is drained by a major river called 'Ala'

Transportation System in Akure Metropolis

Road transportation is the most predominant form of transport in Akure metropolis. It accounts for about 98 percent of all travel in the town. The city of Akure was not planned ab-initio and as a result of this, there is minimal functional relationship between the various land uses. Its poorly developed road network and inefficient traffic management techniques often create chaotic traffic pattern [12] . The traffic composition in the city is mixed comprising of motorcycles, taxis, minibuses, Lorries and trucks (trailers). However, the traffic composition of Akure metropolis is dominated by taxis, motorcycles and buses [13] . The intersection under consideration in this study (Odole Intersection) is cross intersection linking OkeAro (a densely built up area) and Arakale (a commercial nucleus).

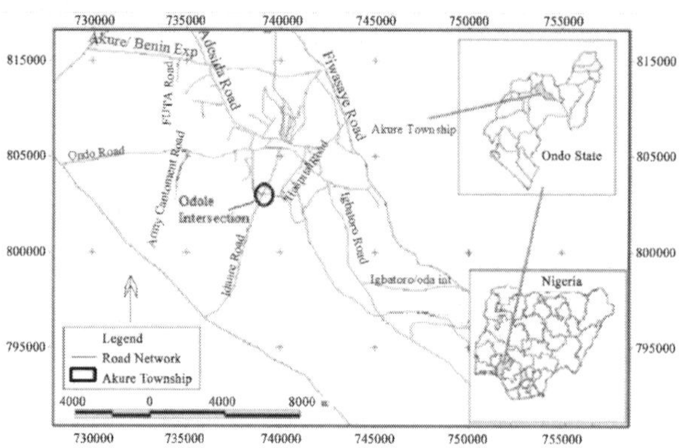

Figure 1. Map of Nigeria showing Ondo state and Akure township road network [14] .

METHODOLOGY

Data Collection

Odole intersection was identified as one of the critical intersections during a reconnaissance Survey and as such, selected for study. Data on geometric features were collected at the intersections using Google Earth software. Geometric details measured include: major road width, minor road width, number of lanes on major road, number of lanes on minor road and angles of inclination of the minor approaches relative to the major route. Data on traffic volume and delay were collected at all approaches during peak (7:30 - 8:30 am) and off peak periods (12.00 noon and 1.00 pm)every other day during weekdays using Cine Camera mounted at a vantage position. Both flow pattern and vehicle classes were monitored. Vehicle classification count was carried out by counting and classifying vehicles as they pass with the aid of tally sheets while flow patterns were documented with due consideration given to diverging, merging and crossing movements to and from each approach using tally sheets while playing back the recorded video. Transportation modes considered include Bicycle, Motorcycle, Passenger Car, Bus, lorry, 2 Axle Truck, 3-Axle Truck, and 4-Axle Truck. The counts were broken down into four flow regimes of 15 minutes interval.

Furthermore, discharge headway data were collected at Odole Intersection using a split timer. It involved recording the successive times it took for vehicles to pass the stop point at the intersection when they are given the right of way. Delay was measured using stop watch by successively taking note of waiting time of the first vehicle that reaches an intersection at an occurrence of stoppage before having the right of way.

Data Analyses

The geometric and traffic data collected were analyzed using Microsoft Excel. The traffic and geometric data collected were utilized in appraising the intersection. Data on geometric features of each intersection were employed in identifying intrinsic geometric problems by comparing the data with standards. Vehicle classification and delay data were used in determination of the proportions of various classes of vehicle plying the intersections under study and Level of Service at the intersection respectively. Saturation headway and discharge lost time were obtained from discharge headway data. To obtain the critical flow pattern at the intersection, the critical hourly volume for each direction of movement in each approach was determined by obtaining the highest volume observed in each regime of flow (15 minutes) and multiplying by 4. Appropriate traffic control mechanism for Odole intersection was selected by juxtaposing the results of geometric and traffic data analyses with the pros and cons of various traffic control mechanisms. This process led to the choice of traffic control by signalisation.

Design of Traffic Signal

Peak and off peak traffic signal designs were carried out to cater for the appreciable difference in the peak and off-peak traffic flow patterns. The choice of the number of phases required for signalization was premised on the existing road geometry and volume of through traffic relative to turning movements of the critical lane. Critical lane volume for each phase was identified by considering the movement with highest traffic volume on the flow pattern diagram. Other parameter required for signal design is Saturation Flow which was calculated as the reciprocal of the Saturation Headway obtained from traffic studies. Finally, Cycle Time and Effective Green Time and Effective Red time were evaluated using appropriate equations.

The mathematical relationships for the determination of cycle length and apportioning of green time are as given thus:

$$\text{Saturation Flow Rate}(s) = \frac{3600}{h_s} \quad (1)$$

where hs is the saturation headway

$$\text{Effective Green Time}(g_i) = G_i + Y_i - t_{Li} \quad (2)$$

where:

Gi = Actual green time for phase i.

Yi = Duration of the yellow interval.

tLi = Total lost time during phase i.

$$\text{Cycle Length}(C) = \frac{Nt_L}{1 - \frac{\sum V_c}{S}} \quad (3)$$

where:

N = Number of phases.

$\sum V_c$= Sum of critical lane volumes.

$$\text{Effective Green time for Phase }(x)(g_x) = \left[\frac{V_{cx}}{\sum_{i=1}^{N} V_{ci}}\right] \times t_g \quad (4)$$

where:

Vcx = Critical lane volume for phase “x”.

$\sum_{i=1}^{N} V_{ci}$= Sum of critical lane volumes.

RESULTS AND DISCUSSION

Geometric Configuration of Odole Intersection

Geometric features of Odole Intersection are presented in Table 1. The major road links are Arakale (North) to OkeAro (South). The minor road links are Gbangbalogun (East) to OdiOlowo (West). There are no inadequacies in the geometric features of the intersection, more so that the intersection is located on a relatively flat terrain.

Analysis of Traffic Composition of Odole Intersection

Figure 2 presents the composition of vehicles observed during peak and off peak traffic counts undertaken at Odole intersection and indicate that the major contributors to traffic at the intersection are Motorcycles (70.88%) and Passenger Cars (28.72%). Other modes of transportation account for about 0.4% of vehicles traversing the intersection.

Analysis of Traffic Volume and Delay at Odole Intersection

Table 2 and Table 3 show the descriptive statistics of traffic volume and delay data collected at the intersection respectively. The delay at Odole intersection is similar at all approaches. The average delay of (22 seconds) is an indication of Level of Service of C which signifies stable operating conditions with average traffic delays at the intersection. This can be attributed to traffic control by personnel at the intersection. However, prolonged traffic jams are usually experienced when traffic personnel are not present.

Table 1. Geometric features of Odole intersection.

Parameter	Value
Major Road Width	9.55 m (Shoulder Inclusive)
Minor Road Width	9.55 m (Shoulder Inclusive)
Number of Lanes on Major Road	2 Lanes (2-way)
Number of Lanes on Minor Road	2 Lanes (2-way)
Angle of Inclination of Minor Approaches Relative to Major Route	≈90°

Table 2. Descriptive statistics of traffic volume at Odole intersection.

Data Type	Maxi-mum	Minimum	Range	Mean	Median	Skew-ness
Peak Traffic Volume (Veh/Hr)	4471	4024	447	4190	4074	1.7
Peak Traffic Volume (Pcu/Hr)	2846	2664	182	2732	2687	1.6
Off-Peak Traffic Volume (Veh/Hr)	4003	3557	446	3797	3830	−0.7
Off-Peak Traffic Volume (Pcu/Hr)	2633	2292	341	2473	2494	−0.5

Table 3. Descriptive statistics of delay at Odole intersection.

Approach	Maximum (secs)	Minimum (secs)	Range (secs)	Mean (secs)	Median (secs)	Skew-ness
Northern	72	2	70	23	20.5	1
Southern	65	3	62	20	16	1.5
Eastern	62	2	60	20	15	1
Western	61	4	57	25	23	0.4
All	72	2	70	22	19	0.9

Secs = seconds.

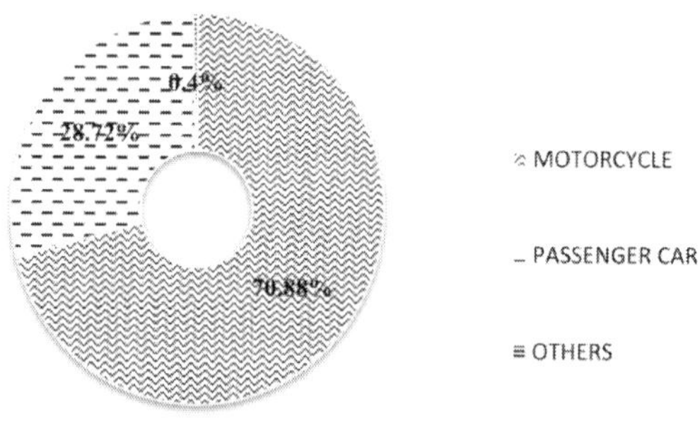

Figure 2. Traffic composition of Odole intersection.

The critical traffic flow patterns at Odole intersection during peak and off peak periods are illustrated by Figure 3 and Figure 4 respectively.

Discharge Headway Results at Odole Intersection

Table 4 and Figure 5 present the Summary of the results of discharge headway study carried out at Odole Intersection. The results reveal that the Saturation Headway (hs) at the intersection is about 1.7 seconds.

Justification for Adoption of Traffic Signalization at Odole Intersection

Free flow of traffic at Odole intersection would remain dependent on the availability of a Traffic Personnel if the regulatory mechanism is left in status-quo. Moreover, a rotary intersection would not avail at Odole Intersection because over 4000 vehicles traverse the intersection in an hour during peak periods. The maximum traffic volume a rotary can efficiently handle is 3000 vehicles per hour [2] . An interchange is unwarranted, considering the cost of construction which surpasses in a great deal, the economic benefits that could be derived from it. From the foregoing, the best alternative suitable for this intersection is signalization since the route is not a high speed route where traffic signalization could cause rear-end-collisions.

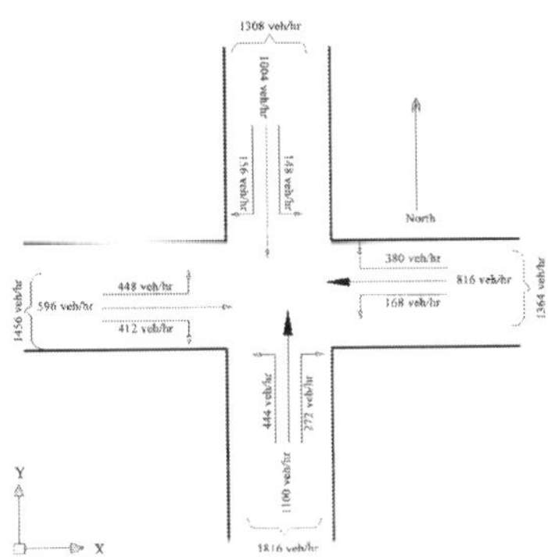

Figure 3. Critical traffic flow pattern at Odole Intersection during peak periods.

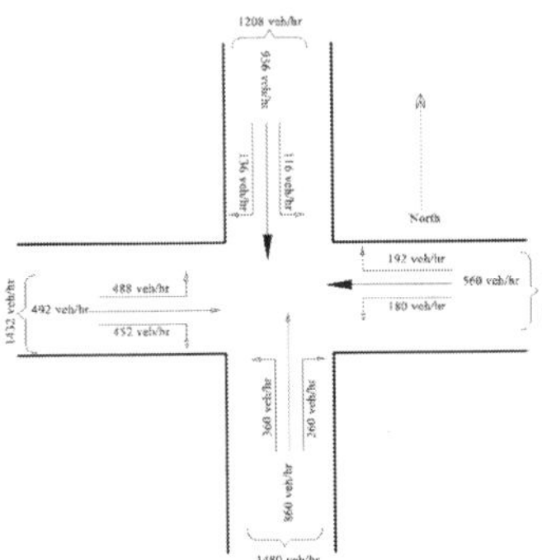

Figure 4. Critical traffic flow pattern at Odole intersection during off-peak periods.

Table 4. Average discharge headway at different vehicle positions at Odole intersection.

Vehicle Position	Average Discharge Headway (h_i) (secs)	Lost Time (l_i) = $h_i - h_s$
1	3.22	1.52
2	2.65	0.95
3	2.27	0.57
4	2.14	0.44
5	1.83	0.13
6	1.82	0.12
7	1.70	0
8	1.70	0
		Total (t_L): 3.73 secs.

Pre-timed Traffic Signal Design for Odole Intersection

Most pre-timed traffic signal designs are geared to the peak time, resulting in frustrating delays to vehicles during off peak periods [8] . Therefore, from the traffic flow patterns, both peak and off peak traffic signal programmes need to be developed because there is an appreciable difference in the volume of traffic during the 2 periods. Furthermore, 2 phase signalization system is the preferred choice for both peak and off peak signal programmes because in both cases, the volume of through traffic is significantly greater than the turning movements. The critical lane volumes are indicated by bold arrows (Figure 2 and Figure 3).

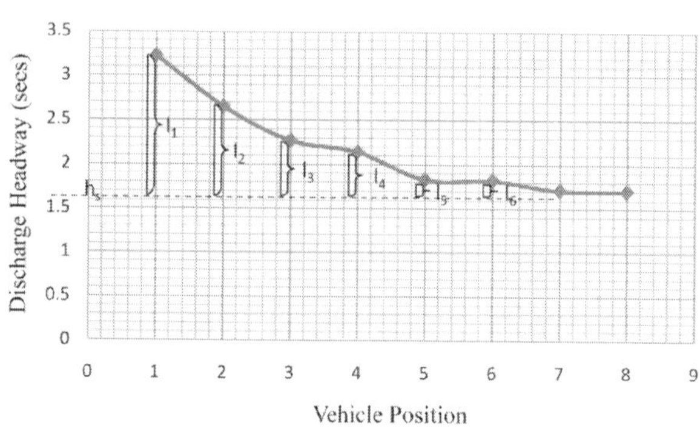

Figure 5. Graphical representation of discharge headway data at Odole intersection.

The traffic studies undertaken at Odole revealed the following signal design parameters:

Average Start-up lost time (tL) = 3.73 secs.

Average Saturation headway (hs) = 1.7 secs.

Sum of critical lane volumes $\left(\sum V_c\right)$ (Peak Period) = 1100 + 816 = 1916 veh/hr.

Sum of critical lane volumes $\left(\sum V_c\right)$ (Off-Peak Period) = 956 + 560 = 1516 veh/hr.

$$\text{Saturation Flow Rate } (S) = \frac{3600}{h_s} = \frac{3600}{1.7} = 2118 \text{ veh/hr}$$

Number of phases (N) = 2 (for both Peak and Off-Peak).

Amber time for Phase 1 (A1): 4 secs.

Amber time for Phase 2 (A2): 3 secs.

where:

Phase 1 gives right of way to major approaches (OkeAro road).

Phase 2 gives right of way to minor approaches (Gbangbalogun-Odiolowo road).

Pre-timed Traffic Signal Design for Peak Period

Since the Saturation Flow Rate is greater than the sum of critical lane volumes, the design can proceed without giving consideration to splitting traffic into more lanes in a bid to reduce sum of critical lane volumes.

$$\text{Cycle Length}(C) = \frac{Nt_L}{1 - \frac{\sum V_c}{S}} = \frac{2 \times 3.73}{1 - \frac{1916}{2118}} = 78 \text{ Secs.}$$

$$\text{Total Effective Green Time}\left(t_g\right) = C - Nt_L = 78 - (2 \times 3.73) = 70.54 \text{ Secs.}$$

$$\text{Effective Green time for Phase } 1\left(g_1\right) = \left[\frac{V_{c1}}{\sum_{i=1}^{N} V_{ci}}\right] \times t_g$$

$$g_1 = \left[\frac{1100}{1916}\right] \times 70.54 = 40.5 \text{ secs}$$

$$\text{Actual Green Time for Phase } 1\left(G_1\right) = g_1 - A_1 + t_L = 40.5 - 4 + 3.73 = 40 \text{ secs}(\text{Rounded})$$

Red time for Phase 1 $= C - G_1 - A_1 = 78 - 40 - 4 = 34$ secs

Effective Green time for Phase 2 $(g_2) = \left[\dfrac{V_{c2}}{\sum_{i=1}^{N} V_{ci}}\right] \times t_g$

$$g_2 = \left[\frac{816}{1916}\right] \times 70.54 = 30 \text{ secs}$$

Actual Green Time for Phase 2 $(G_2) = g_2 - A_2 + t_L = 30 - 3 + 3.73 = 31$ secs(R)ounded

Red time for Phase 2 $= C - G_2 - A_2 = 78 - 31 - 3 = 44$ Secs.

Figure 6 is the timing diagram for the traffic signal during peak periods

It can be gleaned from the result of the design that traffic on the OkeAro road has an upper hand in the apportioning of right of way.

Pre-timed Traffic Signal Design for Off-Peak Period

Since the Saturation Flow Rate is greater than the sum of critical lane volumes, the design can proceed without giving consideration to splitting traffic into more lanes in a bid to reduce sum of critical lane volumes.

$$\text{Cycle Length}(C) = \frac{Nt_L}{1 - \dfrac{\sum V_c}{S}} = \frac{2 \times 3.73}{1 - \dfrac{1516}{2118}} = 26 \text{ Secs.}$$

Total Effective Green Tim e$(t_g) = C - Nt_L = 26 - (2 \times 3.73) = 18.54$ Secs.

Effective Green time for Phase 1 $(g_1) = \left[\dfrac{V_{c1}}{\sum_{i=1}^{N} V_{ci}}\right] \times t_g$

$$g_1 = \left[\frac{956}{1516}\right] \times 18.54 = 11.69 \text{ secs}$$

Actual Green Time for Phase 1$(G_1) = g_1 - A_1 + t_L = 11.69 - 4 + 3.73 = 11$ secs(*Rounded*)

Red time for Phase 1 $= C - G_1 - A_1 = 26 - 11 - 4 = 11$ secs

Effective Green time for Phase 2 $(g_2) = \left[\dfrac{V_{c2}}{\sum_{i=1}^{N} V_{ci}}\right] \times t_g$

$$g_2 = \left[\frac{560}{1516}\right] \times 18.54 = 6.8 \text{ secs}$$

Actual Green Time for Phase 2 $(G_2) = g_2 - A_2 + t_L = 6.8 - 3 + 3.73 = 8$ secs(Rounded)

Red time for Phase 2 $= C - G_2 - A_2 = 26 - 8 - 3 = 15$ Secs

Figure 7 is the timing diagram for the traffic signal during off-peak periods.

There is a marked difference in the peak and off peak timing programmes, corroborating the need for both peak and off-peak timing programmes. However, it is apposite to remark that the traffic signal designs need to be revised from time to time in years to come due to the dynamic nature of traffic behaviour.

CONCLUSION

An appraisal of Odole Intersection has indicated that the major contributors to traffic are Motorcycles (70.88%) and Passenger Cars (28.72%). Other modes of transportation account for about 0.4% of vehicles traversing the intersection.

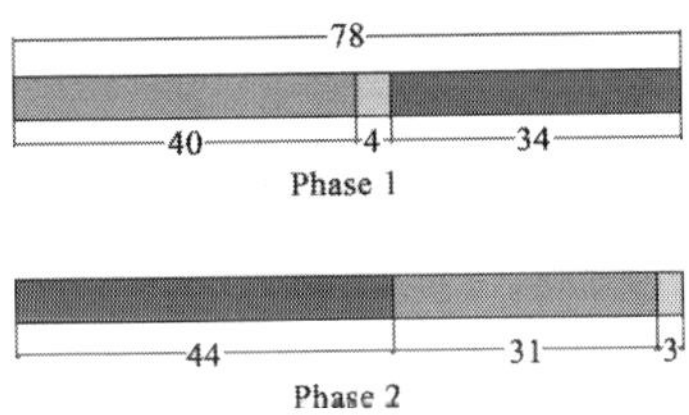

Figure 6. Timing diagram for peak period programme at Odole intersection (all values in seconds).

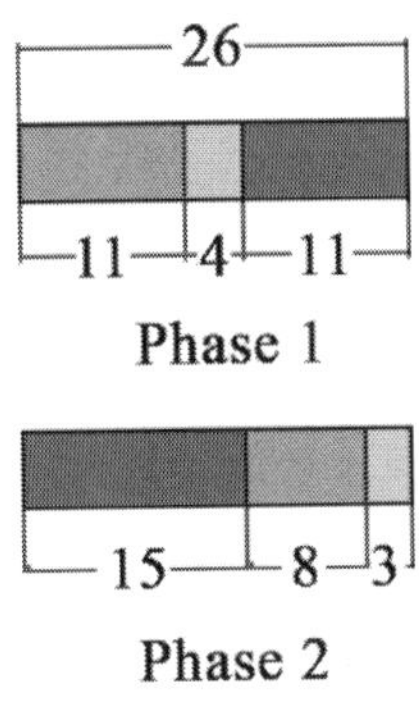

Figure 7. Timing diagram for off-peak period programme at Odole intersection (all values in seconds).

Critical traffic volume at the intersection was over 4000 veh/hr. Furthermore, an average delay of 22 secs at the intersection was an indication of a level of service of C (i.e. fairly stable traffic condition with average delays attributable to traffic control by personnel). Odole intersection has been adjudged to require signalization and has been designed as such, having 40 seconds Green, 4 seconds Amber and 34 seconds Red time intervals in the first phase; 31 seconds Green, 3 seconds Amber and 44 seconds Red time intervals in the second phase for peak period programme. However, off-peak programme has 11seconds Green, 4 seconds Amber and 11 seconds Red time intervals in the first phase; 8 seconds Green, 3 seconds Amber and 15 seconds Red time intervals in the second phase.

RECOMMENDATIONS

In tandem with the findings of this research, the following recommendations are made:

- Construction of Ring Road in Akure Metropolis in order to reduce the burden on existing transportation facilities especially the intersections.
- Dualisation of Oke-Aro road to enable the road to meet up with future traffic demands.
- Reviewing the traffic signal programme at Odole from time to time so as to guarantee its efficiency in controlling any prevailing traffic situation since the designed signal programme would become obsolete over time.

ACKNOWLEDGEMENTS

The authors wish to gratefully acknowledge Prof. A. O. Owolabi for his immense contributions toward the success of this research.

REFERENCES

1. Adams, J. (1981) Transport Planning Vision and Practice. 1st Edition, Routledge and Kegan Paul Ltd., London.
2. Kadiyali, L.R. (2008) Traffic Engineering and Transport Planning. 7th Edition, Khanna Publishers, Delhi.
3. Jain, V., Sharma, A. and Subramanian, L. (2012) Road Traffic Congestion in the Developing World. Proceedings of the 2nd ACM Symposium on Computing for Development, Article 11, ACM New York, New York. http://dx.doi.org/10.1145/2160601.2160616
4. . Engwitch, D. (1992) Towards an Eco-City; Calming the Traffic. Envirobook Publishers, Sussex Inlet.
5. Hoel, L.A., Garber, N.J. and Sadek, A.W. (2008) Transportation Infrastructure Engineering (A Multi-Modal Integration). International Student Edition. Chris Carson, Key Largo.
6. Rogers, M. (2003) Highway Engineering. Wiley-Blackwell Publishing, Ames.
7. Wright, P.H. and Dixon, K.K. (2004) Highway Engineering. 7th Edition, John Wiley and Sons, Hoboken.
8. O'Flaherty, C.A. (1974) Highways Vol. 1 Highways and Traffic. 2nd Edition, Edward Arnold Publishers Ltd., London.
9. Wikipedia (2014) Akure Kingdom. http://wikipedi.org/wiki/Akure_Kingdom
10. Okoko, E.E. (2002) A Predictive Modeling of Spatial Interaction Pattern in the Transport System in Akure. Unpublished PhD Thesis, Federal University of Technology, Akure.
11. Ajibefun, I. (2010) Akure City Profile. International Water Management Institute, Colombo.
12. Owolabi, A.O. (2004) Intra-modal Competition in Urban Para-Transit Systems: Case Study of Akure Metropolis. Unpublished PhD Thesis, Federal University of Technology, Akure.
13. Owolabi, A.O. (2009) Paratransit Modal Choice in Akure, Nigeria. Applications of Behvioural Models. Institution of Transportation Engineers Journal, 79, 54-58.
14. Ayeni, A.O. (2011) Malaria Morbidity in Akure, Southwest, Nigeria: A temporal Observation in a Climate Change Scenario. Trends in Applied Sciences Research, 6, 488-494. http://dx.doi.org/10.3923/tasr.2011.488.494

SECTION II
TECHNIQUES TO REDUCE CRASHES AND INCIDENTS

CHAPTER

5

A Comprehensive Study of Single and Multiple Truck Crashes Using Violation and Crash Data

Mohammad Mahdi Rezapour Mashhadi[1], Shaun S. Wulff[2], Khaled Ksaibati[3]

[1] Department of Civil & Architectural Engineering, University of Wyoming Office: EN 3084 1000 E University Ave, Dept. 3295 Laramie, WY 82071, USA

[2] Department of Statistics, University of Wyoming, Ross Hall 336, Laramie, USA

[3] Wyoming Technology Transfer Center 1000 E. University Avenue Department, 3295 Laramie, WY 82071, USA

ABSTRACT

Around 4,000 people died in crashes involving trucks in 2016 alone in the U.S., with 21 percent of these fatalities involving only single-unit trucks. Many

Citation: M. Mashhadi, Shaun S. Wulff, Khaled Ksaibati, "A Comprehensive Study of Single and Multiple Truck Crashes Using Violation and Crash Data", the Open Transportation Journal, vol. 12, 2018, pp. 43, DOI: 10.2174/1874447801812010043.

studies have identified the underlying factors for truck crashes. However, few studies detected the factors unique to single and multiple crashes, and none have examined these underlying factors in conjunction with violation data. The current research assessed all of these factors using two approaches to improve truck safety. An injury/fatal crash was defined as a crash that results in an injury or fatality. The first approach investigated the contributory factors that increased the odds of injury/fatal single truck and multiple vehicle crashes with involvement of at least one truck. The literature has indicated that previous violations can be used to predict future violations and crashes. Therefore, the second approach used violations related to driver actions that could result in truck crashes. The analysis for the first approach indicated that driving on dry-roadway surfaces, driver distraction, and rollover/jackknife types of truck crashes, speed compliance failure, and higher posted speed limits are some of the factors that increased the odds of injury/fatal single and multiple vehicle crashes. With the second approach, the violations related to risky driver actions, which were underlying causes of truck crashes, were identified and analyses were run to identify the groups at increased risk of truck involved crashes. The results of violations indicated that being nonresident, driving off peak hours, and driving on weekends could increase the risk of truck involved crashes.

Keywords: Single truck crash, Multiple truck crash, Driving violation, Traffic enforcement, Logistic regression, Injury truck crashes.

INTRODUCTION

Trucks are a crucial part of the United States economy. Trucks transport 80% of all freight in the U.S. annually, which accounts for over $700 billion worth of goods [1]. The trucking industry in the U.S. moves about 10.5 billion tons annually, which is expected to increase to 27 billion tons by 2040 [2]. Moreover, seven million people, including more than three million drivers, are employed through this industry. However, truck crashes place a huge burden on the nation in terms of death, injury, and lost productivity. According to the Federal Motor Carrier Safety Administration (FMCSA), there were 667 truck occupant deaths (driver and passenger), and of those 667 deaths, 398 deaths occurred in single-vehicle crashes [3].

Wyoming has the highest fatality rate (24.7 death per 100,000 population) in the nation [4]. Wyoming also has the highest truck crash rate in the United States [5]. These high truck crash and fatality rates result from the high amount of through truck traffic on Wyoming interstates, adverse weather conditions, and mountainous geometric conditions.

However, truck crashes can be mitigated by improving truck safety through policies and regulations, which enhance the performance of the trucking industry without compromising safety. Various countermeasures have been taken in the United States organized into 4 E's of safety. The 4 E's include enforcement, education, engineering, and emergency response. Enforcement is one of the 4 E's that can improve traffic safety. The performance of Wyoming highway patrol (WHP), and consequently road safety, could be improved by identification of the factors that increase the odds of future violations, and consequently future crashes [6, 7]. Thus, this study incorporates violation data, in addition to crash data, to identify the contributory factors to the violations that are likely to increase the odds of future crashes. Identification of these factors can help the WHP to put more emphasis on the contributory factors of risky violations resulting in traffic safety.

Truck crashes are complex events. They can involve single vehicles or two or more vehicles. Out of 700 truck occupant deaths that occur every year in the U.S., about 60% occur in single-vehicle truck crashes [8]. For each type of event, different contributory factors may play roles. Previous research indicated that there are significant differences between single and multiple vehicle crashes [9, 10]. Therefore, this study analyzed single truck and multiple vehicle crashes, with truck involvement, separately. This study investigated factors impacting different types of truck crashes by including vehicle, driver, and environmental factors. In addition, this study included violation data to identify the groups at higher risk of truck crashes by including only the violations contributing to truck crashes in this state. For the purpose of this study, a truck is defined as a commercial vehicle with gross vehicle weight rating greater than 10,000 pounds.

BACKGROUND

Based on FMCSA, the critical reasons for large truck crashes can be assigned to driver (87%), non-performance (12%), recognition (28%), decision (38%), performance (9%), and vehicle (10%) [11]. Lemp *et al.* (2011) used the ordered probit model to investigate the impact of vehicle, occupant, driver, and environmental characteristics on crash severity for those involved in heavy-duty truck crashes [12]. The results indicated that the odds of fatalities increase with the number of trailers and fall as the truck gross vehicle weight rating decreases. Khattak *et al.* (2003) used crash data in North Carolina during 1996-1998 to investigate the impact of truck rollovers and occupant injuries in single-vehicle crashes [8]. The results indicated that higher risk factors in single-truck-crashes include risky driving, speeding, alcohol and drug use, traffic control violations, truck exposure to dangerous road geometry, and trucks that transport hazardous materials.

Moomen *et al.* (2018) investigated the influential factors of downgrade truck crashes in Wyoming using logistics regression. They found that driver gender, speed compliance, weather, lighting and road condition, shoulder and lane width, number of sag and crest curves, roadway grade and length are the contributory factors to truck related crashes [13]. Schneider *et al.* (2009) developed multinomial logit models to investigate driver injury severity resulting from single-vehicle crashes [14]. Different driver, vehicle, and environmental characteristics were found to increase injury severity. Being female, older, unbuckled, fatigued, and under the influence led to increase in the odds of injury. Zhu and Srinivasan (2011) investigated the factors impacting the injury severity in truck crashes [15]. Truck driver distraction, alcohol use, and emotional factors of car drivers were associated with higher severity crashes. Pahukula *et al.* (2015) investigated the contributory factors to injury severity of truck crashes using data from Texas during 2006 to 2010 [16]. The results indicated that different time periods in a day have different contributing impacts on truck crash severity.

However, in the majority of the studies, researchers mostly looked at the injury severity of both multiple-vehicle and single truck crashes as a whole. Thus, they did not identify the variables unique to single and multiple crashes. Zou *et al.* (2017) carried out a study in New York City to investigate the differences between single-vehicle and multiple-vehicle truck crashes [17]. The results indicated that there are substantial differences between factors affecting single and multiple truck crashes. Thus, this study examined truck crash severity separately for single-vehicle and multiple-vehicle truck crashes.

Many studies have identified correlations between previous violations and future crash risk. A previous study carried out by Li and Baker (1994) indicated that conviction records can be used to identify groups with greater odds of involvement in fatal crashes [18]. Similarly, Elliott (2001) investigated the ability of previous violations to predict future offenses and crashes [19]. The results indicated that the drivers with previous ticketed offenses are at greater risk for future crashes. Rezapour *et al.* (2017) used violation data, in addition to crash data, to assess unsafe driver actions to reduce crashes [6]. Chen *et al.* (1995) carried out a study by examining driver records to investigate the relationship between crashes and past records of crashes and convictions [20]. The authors found a positive correlation between pre-period crashes per driver and pre-period number of convictions. In this study, failure to yield and disobeying traffic signals were two violations that best predict crashes. Lantz and Loftus (2006) carried out a study to use an analytical model for predicting future crash involvement based on the history of driver information and also identifying effective enforcement actions that can predict driver behavior and future crash involvement [21]. The results indicated that reckless driving and improper turn violations are the violations that have the highest increase in likelihood of a future crash. Also, failure to keep proper lane was some of the convictions with the highest likelihood of a future crash. A study by Terrill *et al.* (2016) investigated the impact of traffic citations on the number of crashes on an interstate in Wyoming [22]. The results indicated that an increased number of citations issued is a preventive measure for the number of crashes.

However, none of the aforementioned studies used violation or conviction data to investigate groups of truck drivers with an increased risk of being involved in truck crashes. On the basis of the discussed studies, violations can be used as an indication of the groups that are at greater risk of being involved in future crashes.

This current study was set forward to fulfill two main objectives:

1. Conduct crash analysis to determine the factors impacting injury single truck and multiple vehicle, truck involved, crashes. In order to determine these factors, two analyses were carried out:
    1. 1.1 Injury/fatal single truck crash analysis.
    2. 1.2 Injury/fatal multiple vehicle crash analysis with involvement of at least one truck.
2. Conduct violations analysis to identify the groups who are more likely to violate the laws that are the main causes of single and multiple truck crashes. Two analyses were carried out to fulfill this objective:
    1. 2.1 Analysis of those types of violations associated with single truck crashes.
    2. 2.2 Analysis of those types of violations associated with multiple vehicle crashes involving at least one truck.

A crash in this study is one that results in an injury or fatality. Due to the low number of fatality crashes, these crashes were aggregated with injury crashes.

METHODS

Logistic regression is used in many studies involving binary crash outcomes [23-27]. For the logistic regression model, the binary response variable *Y* is assumed to have a Bernoulli distribution with probability π [28].

$$\log \frac{\pi}{1-\pi} = \underline{x}'\underline{\beta} \tag{1}$$

where *x* is a vector of explanatory variables and $\underline{\beta}$ is a vector of unknown regression coefficients. Equation (**1**) can be solved for π which gives

$$\pi = P(Y = 1) = \frac{\exp(\underline{x}'\underline{\beta})}{1 + \exp(\underline{x}'\underline{\beta})} \tag{2}$$

For modeling truck crash severity, the binary response variable was 1 for injury/fatal and 0 for a property damage only (PDO) truck involved crashes. The response is conditioned on a crash that has occurred, and then looking at its binary classification (fatal/injury (F+I) or PDO). Separate models were developed for single truck crashes and for multiple vehicle, truck involved, crashes. The probability (π) of either a single or multiple truck crash being injury was modeled using various risk factors as explanatory variables ($\underline{x}$).

Logistic regression was also used for analyzing violations. The purpose of these analyses was to identify drivers who are more at risk of committing particular traffic law violations, which can lead to truck involved crashes. Here, the response (*Y*) had the value 1 if a driver received a citation of a particular type and a 0 if a driver did not receive a citation of a particular type. The response is conditioned on drivers who had received a violation and then looking at its binary classification for the citation type (received or did not receive citation of particular type). Two different citation analyses were considered. The first analysis involved only those violations more commonly observed with single-truck crashes. The second analysis involved those violations most commonly observed in multiple-vehicle crashes involving at least one truck. The probability (π) of a driver receiving a citation of a particular type was examined in relation to the explanatory variables ($\underline{x}$) involving driver characteristics such as gender and residency, and temporal characteristics such as time of day and day of the week.

Stepwise model selection was used to select explanatory variables for a final logistic regression model. A significance level of 0.10 was pre-specified for entering the model and a significance level of 0.05 was pre-specified for staying in the model. All analyses were performed using the Statistical Analysis System (SAS) [29].

DATA PREPARATION

The data was combined from the three interstates in Wyoming, I-80, I-25, and I-90, with the highest truck related crash rates. Crash data was obtained from the Wyoming Department of Transportation (WYDOT) using the Criticial Analysis Reporting Environment (CARE) from 2011 to 2014. This study used various variables, which can be categorized under driver, environmental, vehicle, temporal, crash, and driver behaviors. Driver characteristics included age, gender, residency, violation (conviction) record, and speed limit compliance at the time of crash. Environmental characteristics included weather and roadway-surface conditions. Weight of a truck was categorized as a vehicle characteristic. Day of week and time of crash were organized under temporal characteristics. Roadway characteristics included posted speed limit of a location where a crash occurred. Driver actions at the time of crash, number of vehicles, and pre-collision vehicle action were categorized under crash characteristics. Driver distraction, driver under influence (DUI) suspicion, fatigue, and the use of safety technology were categorized under driver behaviors. In this study, distraction is defined as any type of distraction such as TV, cell pager, or wireless communication inside the cabin at the time of crashes. Truck crash analyses were divided into two parts: Single truck and multiple vehicles, truck involved, crashes. Single truck crashes were investigated separately as more than 50% of all the truck crashes were single truck crashes.

The violation data was obtained from the Wyoming court reported violation database from 2011 to 2014. For single truck crash analysis, truck drivers were at fault in the crashes. Therefore, the violation data for this analysis was filtered to include just truck driver violations to identify groups that are more at risk of single truck crashes.

There were 121,680 violations filtered to 17,239 truck violations. However, all violations were used for investigating the groups that were at higher risk of multiple vehicles crashes, involving at least one truck. This is due to the fact that both truck and no truck drivers could be at fault in these crashes. Only violation types: follow too closely, failure to drive within single lane, and speed too fast for conditions that resulted in truck crashes were included in this study (Table **1**). Violations related to the main causes of truck crashes were identified among 800 types of violations and presented in Table **1**. For instance, only one violation type: driving too fast for conditions was identified as a contributing factor to a crash type: "drove too fast for conditions".

Table 1. Driver Actions and related Violations, truck involved crashes.

Driver Actions, Truck Crashes	Number of Identified Violation	Driver Actions, Related Violation
Failed to keep Proper Lane	1	Fail to drive within single lane
Drove too fast for conditions	1	Speed too fast for condition
Followed too close	1	Follow too closely

RESULTS

Descriptive Analysis, Crash Data

Fig. (**1**) presents general characteristics of truck crashes on Wyoming interstates. As can be seen from Fig. (**1a**), most of the truck crashes (52%) involved just a single truck. Including both single (52%) and multiple vehicle, truck involved crashes (26%), about 78% of the truck drivers were at fault for truck crashes on Wyoming interstates. Driver actions with highest percentage, for both single and multiple truck related crashes, are included in Figs. (**1b** and **c**).

As can be seen from Figs. (**1b** and **c**), the main causes of truck crashes, driver actions, include failing to keep proper lane, driving too fast for conditions, and following too close. Based on Figs. (**1b** and **c**), no improper driving, failure to keep proper lane, following too close account for 78% of multiple vehicle, truck involved, crashes and 65% of single truck crashes. That is the reason why the violations related to these driver actions are included in the violation analyses (Table **1**).

It should be noted that "no improper driving" action is a crash in which a driver had no improper driving, but was involved in a crash.

Therefore, no violation was identified related to this driver action and this driver action was excluded from violation analyses.

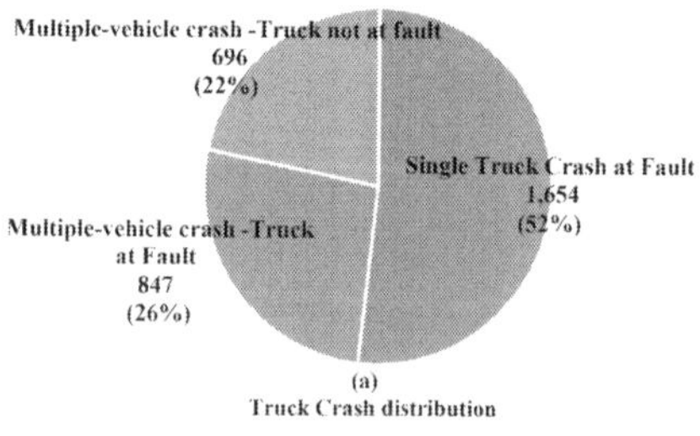

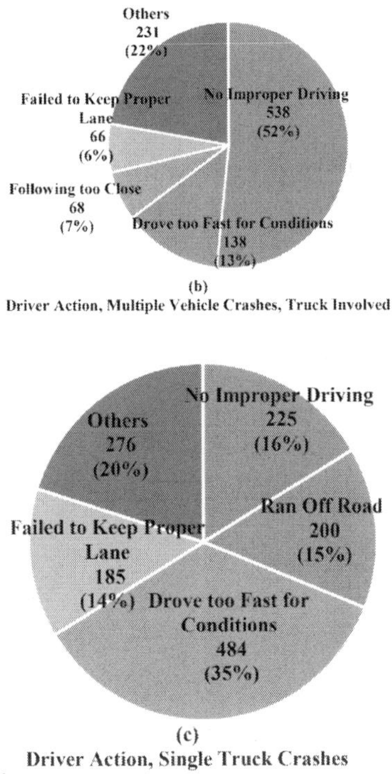

Figure 1. Different characteristics of truck crashes, 2011-2014.

For the first analysis, data were filtered from the original file to include only single truck crashes. Summary statistics of significant variables that impacted the severity of single truck crashes are presented in Table 2. As can be seen from the table, most of the single truck crashes (85%) involved property damage only. Most of the drivers in single truck crashes were male (94%) compared with only 6% involving female drivers. Speed limit of 65 mi/hr was chosen as a threshold for the speed limit variable. This is because most speed limits on the included highways in Wyoming are greater than 70 mi/hr and 65 mi/hr was found to be the best threshold that can divide the crashes into similar categories. Although the majority of single truck crashes (71%) occurred at a posted speed limit of greater than 65 mi/hr, a rather large proportion (29%) occurred at a lower speed limit. Most of the single truck crashes (67%) occurred on not-dry-road conditions. Most single truck crashes (72%) were rollover or jackknife. In 16% of all the single truck crash cases, truck drivers had some type of distraction in the cabin. Distraction was defined as any distraction in a cabin such as wireless communication or TV.

Table 2. Summary statistics of single and multiple truck crashes, Wyoming Interstates.

Truck crashes			
Variable Name		**Number**	**%**
Single Truck Crashes			
Injury single truck crash	Injury truck crash	204	15
	Property damage only	1166	85
Gender	Male	1290	94
	Female	75	6
Posted speed limit	Less than or Equal 65 MI/HR	396	29
	Greater than 65 MI/HR	972	71
Road condition	Dry	450	33
	Not dry	920	67
Rollover/Jackknife	Rollover/Jackknife	993	72
	Other types	377	28
Driver distraction	Has some Types of Distraction in Cabin	217	16
	There was no distraction	1139	84
Multiple Vehicle Crashes, Truck Involved			
Injury multiple vehicle crash, truck involved	Injury truck crash	85	7
	Property damage only	1,203	93
Speed compliance	Speed limit was followed	1,203	93
	Speed limit was not followed	85	7
Speed Limit was not followed	Dry-road condition	32	38
	Not-dry-road condition	53	62
Posted Speed Limit	Less than or equal 65 mi/hr	365	29
	Greater than 65	895	71
Lower posted speed Limit	Dry-road condition	161	44
	Not-dry-road condition	259	56
Crash characteristic	Truck was at fault of crash	749	58
	Non-truck driver was at fault of the crash	539	42
Temporal characteristics	Business day	808	63
	Weekend	480	37

For the second analysis, the data was filtered from the original file to include crashes involving at least one truck. Due to the involvement of at least two vehicles, only the summary statistics of a vehicle at fault is presented in Table 2. Seven percent of truck crashes occurred while the driver at fault did not follow the posted speed limit. To provide more insight about this variable, Table **2** also includes more statistics on the circumstance in which speed limit compliance was not fulfilled. In these crashes, 62% of the at-fault vehicles did not follow the posted speed limit while driving on not-dry-road conditions. Not-dry-road conditions include the road conditions other than dry, such as rainy/snowy. About 29% of truck crashes occurred at the locations with a speed limit of less than 65 mi/hr. More detailed summary is also provided for this variable in Table **2**. The results indicated that most of the lower speed crashes occurred when the road was not dry (56%), which might be an indication that the locations were equipped with Variable Speed Limits (VSL). Forty four percent of the lower speed limits were also related to dry-road condition, which might be due to driving through work zones.

Statistical Modeling, Crash Data

Factors Associated with Higher Risk of Injury Single Truck Crashes

This first modeling approach investigated the variables that increase the odds of fatal/injury truck crashes compared to PDO truck crashes. Table **3** shows the variables included in the full model and the estimates for those variables remaining in the reduced model at the pre-specified significance levels. Being a male driver decreased the odds of being involved in injury single truck crashes; females are not more likely to be involved in a crash. However, when they are involved in a crash, they are more likely to be injured/killed. The results disagreed with the research carried out by Kim *et al.* (2013) indicating that being a male truck driver increases the odds of F+I single truck crashes [30]. It was found that dry-road conditions increase the odds of being involved in injury single truck crashes. This result is in contrast with the results obtained by Kim *et al.* (2013) indicating that wet or snowy/icy surfaces increased the odds of injury in single-vehicle crashes [31]. However, the difference between single vehicle and single truck crashes should be noted.

The lower odds of injury single truck crashes may lie in the fact that truck drivers drive more cautiously, with lower speed, on not-dry-road conditions. Speed limit is another variable that impacts the severity of single truck crashes on Wyoming interstates. The results indicated that by increasing speed limit, single truck crashes were more likely to be injury/fatality. The results confirmed the research carried out by Abdel-Aty (2000), which indicated that when drivers speed, the odds of being involved in an injury crash increase [32]. Rollover/jackknife was another important variable. The odds of an injury/fatal truck crash were estimated to be nearly 4 times higher if that truck crash involved a rollover/jackknife than if that truck crash did not involve a rollover/jackknife. The result was in accordance with the result found by Krull *et al.* (2000) [33]. It should be noted that this research included only rollover in all types of single vehicle crashes. Driver distraction was found to increase the odds of getting involved in injury single truck crashes. The result confirmed the research carried out by Bunn (2005), which indicated distraction/inattention increased the odds of a fatal motor vehicle collision [34].

Table 3. Contributory factors to injury single truck crashes, Wyoming Interstates.

Collision Characteristic	**Predictors**	**Odds Ratio**	**95% CL**
Driver characteristics	**Gender:** Male truck driver (1 if true; 0 otherwise)	0.35*	0.19-0.62
	Age group[1]: Younger driver: age <45 years (1 if true; 0 otherwise)	-	-
	violation record[1] : Traffic record of one ticket or less (0 if true; 1 otherwise)	-	-
	Residency[1]: State of Wyoming (0 if true; 1 otherwise)	-	-
	Speed limit compliance[1]: Truck Driver followed Speed Limit(0 if rue, 1 otherwise)	-	-
Environmental characteristics	**Weather condition[1]:** Clear (0 if true; 1 otherwise)	-	-
	Roadway-surface condition: Dry (0 if true; 1 otherwise)	0.19*	0.14-0.27
	Light condition[1]: daylight (0 if true; 1 otherwise)	-	-
Vehicle characteristics	**CMW weight[1]:** Truck greater than 26000 pound (1 if true; 0 otherwise)	-	-
Temporal characteristics	**Day of week[1] :** Weekends(1 if true; 0 otherwise)	-	-
	Time[1]: Peak hours 6< <22: (0 if true; 1otherwise)	-	-

Collision Character-istic	Predictors	Odds Ratio	95% CL
	Posted speed limit : Speed limit less than 65 mi/hr(0 if true; 1 otherwise)	1.63*	1.11-2.47
Roadway characteristics	**Traffic**[1]	-	-
Crash characteristics	**No of vehicle** [1]**:** 1(0 if true; 1 otherwise)	-	-
	Pre-collision vehicle actions[1]**:** Straight-ahead(0 if true; 1 otherwise)	-	-
	Manner of collision[1]	-	-
	Rollover/Jackknife(1 if true; 0 otherwise)	3.62*	2.38-5.52
	Occupant ejection[1]**:** Driver is partially or totally ejected(1 if true; 0 otherwise)	-	-
	No improper driving[1] (1 if true; 0 otherwise)	-	-
	Drove too fast for conditions[1]**:** (1 if true; 0 otherwise)	-	-
	Fail to keep proper lane[1]**:** (1 if true; 0 otherwise)	-	-
Driver behaviors	**Driver distraction:** No distraction in truck(0 if true; 1 otherwise)	1.72*	1.16-2.60
	DUI suspicion[1] **:** Driver was suspected of driving under the influence condition(1 if true; 0 otherwise)	-	-
	Sign of fatigue[1]**:** Driver was fatigued(1 if true; 0 otherwise)	-	-
	CMV driver safety technology in use[1]**:** Safety equipment was not used (1 if true; 0 otherwise)	-	-

Severity (response): no injury=0, injury or fatality=1

*indicates $p<0.05$

1 indicates insignificant variable included in the initial model

Table **4** provides possible explanations behind the variables identified in the logistic model for severity of single truck crashes. For instance, the increased odds of being involved in injury/fatal single truck crashes, with distraction in the cabin, may be related to drivers being less attentive to road hazards. Also, a female driver may have higher odds of being involved in injury/fatal single truck crashes since females are less able to sustain different types of physical trauma [35].

Table 4. Significant variables and the possible reasons for the signs, single truck crashes, Wyoming Interstates.

Variable	Relationship	Possible Reason
Female truck driver	It increases the odds of injury/ fatal single truck crashes.	Women are less able to sustain physical trauma.
Wet roadway surface condition	It decreases the odds of injury/ fatal single truck crashes.	Drive more cautiously, with less speed, on not-dry-road conditions.
Increased posted speed limit	It increases the odds of injury/ fatal single truck crashes.	Drive faster in higher posted speed limit locations.
Rollover/Jackknife	It increases the odds of injury/ fatal single truck crashes.	Increased the contact of driver with road and vehicle.
Driver distraction	It increases the odds of injury/ fatal single truck crashes.	Drivers are less cautious about the hazard associated with Wyoming mountainous area.

Factors Associated with Higher Risk of Injury Multiple Vehicle Crashes, Truck Involved

After learning about contributory factors to injury single truck crashes, an analysis was run to identify contributory factors to injury in multiple vehicle crashes, with involvement of at least one truck. Out of 25 included variables, 3 variables were found to be important at the pre-specified significance level. As can be seen from Table **5**, lack of speed limit compliance had the highest estimated impact among the variables in this crash analysis. The odds of this crash were estimated to be 3.34 times higher when the truck driver did not follow the speed limit. Although not much research has been done on the importance of speed compliance in preventing injury truck crashes, many studies showed that increased vehicle speed significantly increased the odds of being involved in fatal/injury crashes [36, 37]. Posted speed limit is another speed variable that increased the odds of being involved in fatal/injury crashes. An increase in injury/fatal crashes was identified during weekends. This impact can be due to a variety of reasons such as less congestion traffic, higher travel speeds, and a desire to reach a destination in a timely manner. The possible reasons for other important variables are presented in Table **6**.

As can be noticed from the results presented in Tables **3** and **5**, the only common predictor between crash severity for single truck and multiple vehicle, truck involved, crashes related to posted speed limit. Increased posted speed limit increased the crash severity for single truck crashes (1.63) and multiple vehicle crashes (1.85).

Table 5. Contributory factors to injury multiple vehicle crashes, Wyoming Interstates.

Collision Characteristic	Predictors	Odds Ratio	95% CL
Driver characteristics	**Gender[1]:** Male truck driver (1 if true; 0 otherwise)	-	-
	Age group[1]: Younger driver: age <45 years (1 if true; 0 otherwise)	-	-
	violation record[1] : Traffic record of one ticket or less (0 if true; 1 otherwise)	-	-
	Residency[1]: State of Wyoming (0 if true; 1 otherwise)	-	-
	Speed Limit compliance[1]: Truck Driver followed Speed Limit(0 if rue, 1 otherwise)	3.34*	1.75-6.35
Environmental characteristics	**Weather condition[1]:** Clear (0 if true; 1 otherwise)	-	-
	Roadway-surface condition[1]: Dry (0 if true; 1 otherwise)	-	-
	Light condition[1]: daylight (0 if true; 1 otherwise)	-	-
Vehicle characteristics	**CMW weight[1]:** Truck greater than 26000 pound (1 if true; 0 otherwise)	-	-
Temporal characteristics	**Day of week :** Weekends(1 if true; 0 otherwise)	1.70*	1.22-2.36
	Time[1]: Peak hours 6< <22: (0 if true; 1otherwise)		
	Posted speed limit : Speed limit less than 65(0 if true; 1 otherwise)	1.85*	1.25-2.74
Roadway characteristics	**Traffic[1]**	-	-
Crash characteristics	**No of vehicle [1]:** 1(0 if true; 1 otherwise)	-	-
	Pre-collision vehicle actions[1]: Straight-ahead(0 if true; 1 otherwise)	-	-
	Manner of collision[1]	-	-
	Rollover/Jackknife[1] (1 if true; 0 otherwise)	-	-
	Occupant ejection[1]: Driver is partially or totally ejected(1 if true; 0 otherwise)	-	-
	No improper driving[1] (1 if true; 0 otherwise)	-	-
	Drove too fast for conditions[1]: (1 if true; 0 otherwise)	-	-
	Fail to keep proper lane[1]: (1 if true; 0 otherwise)	-	-
Driver behaviors	**Driver distraction[1]:** No distraction in truck(0 if true; 1 otherwise)	-	-
	DUI suspicion[1] : Driver was suspected of driving under the influence condition(1 if true; 0 otherwise)	-	-
	Sign of fatigue[1]: Driver was fatigued(1 if true; 0 otherwise)	-	-
	CMV driver safety technology in use[1]: Safety equipment was not used (1 if true; 0 otherwise)	-	-

Table 6. Significant variables and the possible reasons for the signs, multiple vehicle crashes, Wyoming Interstates.

Variable	Relationship	Possible Reason
No speed compliance	It increases the odds of getting involved in injury multiple vehicle crashes	Drive faster than recommended speed limit.
Weekend driving	It increases the odds of getting involved in injury multiple vehicle crashes.	Less traffic, higher speeds, or rush to reach a destination.
Increased posted speed limit	It increases the odds of getting involved in injury multiple vehicle crashes	Drive faster on increased posted speed limit locations.

Descriptive Analysis, Violation Data

The literature indicated that previous violations can be used to predict future crashes. Therefore, for a better understanding of the enforcement efficiency, this section aimed to include the most common types of driver violations resulting in truck crashes. The objective of this section was to identify the groups that are more at risk of violating the laws that can result in truck related crashes. Figs. (**1b** and **c**) showed the highest proportions of driver actions in truck involved crashes. The figures revealed that driving too fast for conditions (13% for multiple vehicle crashes compared to 35% for single truck crashes), no improper driving (16% for single truck crashes compared to 52% for multiple vehicle crashes), failure to keep proper lane (6% for multiple vehicle crashes compared to 14% for single truck crashes), and following too close (7% of multiple vehicle crashes) accounted for 65% of all the single truck crashes and 78% of multiple vehicle crashes. Therefore, violations related to aforementioned driver actions were included in the violation analyses. No violation was identified for "no improper driving" action as drivers had no improper driving, but were involved in this type of crashes. Therefore, this type of driver action was excluded from the violation analyses.

The summary statistics of significant variables and the summary statistics of the responses are included in Table (**7**). Violation analyses were divided into two sections: truck related violations and all types of violations. For single truck crashes, only trucks were at fault in the crashes, so only truck violation data were used to identify truck drivers being at higher risk of future crashes. For multiple vehicle crashes, with an involvement of at least one truck, truck and no truck vehicles/drivers could be at fault in the crashes.

Therefore, all types of violations were used for investigating the drivers at higher risk of involvement in multiple vehicle crashes, with involvement of at least one truck. Only 12% of all the citations were assigned to trucks. Other types of violation, "others", included violations such as seat belt, driving under the influence (DUI), and Hour Of Service (HOS) violations. Nonresidents of Wyoming accounted for 70% of all the truck violators and 33% of all the violators. The WHP allocated only 11% of all their resources during off peak hours and most of the violators were male (97%).

Table 7. Summary statistics of violations on Wyoming Interstates, truck violation.

Violation	Number	%
Truck Violation	17,239	12
Non-truck Violation	121,680	88
Truck Related Violations		
Speed too fast for condition	226	1.38
Fail to drive within single lane	755	4.62
Speeding	3,179	19.5
Others	11,800	70.3
Wyoming Resident	2,010	12
Non Wyoming Resident	14,327	88
Weekend	4,423	27
Business days	11,914	73
Peak Hours	14,584	89
Off Peak Hours	1,753	11
Male	15,814	97
Female	482	3
All Violations		
Speed too fast for condition	1,239	0.9
Fail to dive within single lane	2,930	2.1
Speeding	86,984	62.6
Following too Close	1,144	0.8
Others	45,855	33
Weekend	39,568	30
Business days	93,266	70
Peak Hours	124,621	94
Off Peak Hours	8,113	6.1

Statistical Modeling, Violation Data

Factors Associated with Higher Risk of single Truck Crashes, Risky Truck-Related Violations

This section identified the risk of different truck drivers in violating particular traffic laws, with different driver and temporal characteristics. Only truck related violations were considered in this section as the risk of involvement in single-truck crashes were evaluated. Also, only traffic violation types that are associated with single truck crashes were included in this analysis (Fig. **1b**). Therefore, the truck related violations were filtered to include only violations titled: Speed too fast for conditions and failure to drive within single lane. The aforementioned violations, driver actions, accounted for 49% of all single truck crashes, driving too fast for conditions (35%), and failure to drive within single lane (14%). Results of this analysis are presented in Table **8**. As can be seen from this table, nonresidents of Wyoming, driving on weekend, and driving in off peak hours, all increased the odds of driving trucks too fast for conditions. These variables, based on the literature, could consequently increase the risk of being involved in single truck crashes. In particular, the odds of violating driving too fast for conditions were estimated to be 2.21 times higher for a nonresident than for a resident of Wyoming.

Except for failure to drive within single lane, all the variables included in the model were found to affect the odds of violating this law. Being female, a nonresident of Wyoming, driving on the weekend, and driving off peak hours all increase the odds of a violation related to failure to keep the proper lane. Being nonresident (2.00) and driving during weekends (2.31) have the highest estimated odds of such a violation.

Table 8. Adjusted odds ratio for driver and temporal characteristics, truck related violation on Wyoming interstates, 2011-2014.

–		**Speed Too Fast for Condition**	**Failure to Drive within Single Lane**
Driver characteristics	**Driver gender** Male truck driver (1 if true; 0 otherwise)	-	0.65*
	Residence group: Non residence(1 if true; 0 otherwise)	2.21*	2.00*

–		Speed Too Fast for Condition	Failure to Drive within Single Lane
Temporal characteristics	**Date:** weekends(1 if true; 0 otherwise)	1.97*	2.31*
	Time: peak 6< <22(0 if true; 1 otherwise)	1.55*	1.25*

*indicated $p<0.05$

-indicates insignificant variable

Table 9. Significant variables and the possible reasons for the signs, single truck violations.

Variable	Relationship	Possible Reason	Possible Reason
Female	Increase the risk of not keeping proper lane.	Have higher score for errors and violations.	Have higher score for errors and violations.
Non residence	Increase the risk of driving too fast for the condition and failing to keep proper lane.	Lack of familiarity with mountainous areas in Wyoming.	Lack of familiarity with mountainous areas in Wyoming.
Weekend	Increase the risk of driving too fast and failing to keep proper lane.	Less traffic and higher speeds.	Less traffic and higher speeds.
Off peak hours	Increase the risks of all these violations.	Possible fatigue.	Possible fatigue.

Factors associated with higher risk of multiple vehicle, truck involved crashes, all types of risky violations

As Fig. (**1a**) indicated, in 22% of the multiple vehicle crashes, truck involved, non-truck drivers could also be at fault. As a result, all types of violations, truck and no truck violations, were included in this analysis to identify the risk of different drivers, with different temporal and driver characteristics, in violating particular traffic laws. Only traffic violations that are the main causes of multiple vehicle truck crashes were included in this analysis. As descriptive analysis in Fig. (**1b**) shows, 26% of multiple vehicle crashes were related to driving too fast for conditions (13%), following too close (7%), and failure to keep proper lane (6%). Therefore, only violations related to aforementioned driver actions were included in the violation analysis.

As can be seen from Table **10**, being a truck driver increased the odds of violating all of these laws. This is consistent with the results in Fig. (**1a**) showing that truck drivers are at fault in 78% of all truck related crashes.

Contrary to the results obtained from single truck violation analysis, driving on weekends decreased the odds of driving too fast for conditions and failing to drive within single lane violations. The possible reason for the difference between the impact of this variable for this analysis and single truck violation analysis could be related to the fact that in this analysis about 90% of violators are non- truck drivers. On the other hand, driving off peak hours, similar to the previous analysis, was an indication of increased odds for driving too fast for conditions and failing to keep proper lane.

Table 10. Adjusted odds ratio for driver and temporal characteristics, all types of violation on Wyoming interstates, 2011-2014.

Groups		Drive too Fast for Condition	Failure to Drive within Single Lane	Following too Close
Driver characteristics	**Driver gender** Male truck driver (0 if true; 1 otherwise)	-	-	-
	Residence group: Non residence(1 if true; 0 otherwise)	-	-	-
Type of vehicle	**No truck**(0 if true; 1 otherwise)	2.22*	2.85*	2.26*
Temporal characteristics	**Date:** weekends(1 if true; 0 otherwise)	0.27*	0.60*	-
	Time: peak 6< <22(0 if true; 1 otherwise)	1.67*	3.00*	-

*indicated $p<0.05$

-indicates insignificant variable.

CONCLUSION AND DISCUSSION

Crash data from WYDOT was used to study the impacts of various variables on injury/fatal single truck and multiple vehicle, truck involved crashes. While much research has been done on the contributory factors to both multiple-vehicle and single truck crashes as a whole, this work investigated multiple-vehicle and single truck crashes separately. One of the reasons that single truck crashes were analyzed separately is the importance of this type of crash on Wyoming interstates. As it was shown in this study, more than 50% of all the truck related crashes were related to single truck crashes. This study also just focused on mountainous interstates in Wyoming that have the highest truck crash rates in the state. Moreover, this study included violation data in addition to truck related crash data.

The results indicated that various variables impact single truck and multiple vehicle crashes, with involvement of at least one truck. Being female, driving on dry-road conditions, driving with increased posted speed limits, rollover/jackknife type of truck crashes, and having distraction in the cabin were factors that increased the odds of injury/fatal single truck crashes. As for multiple vehicle crashes, truck involved, different variables impacted this type of crash. Non-speed limit compliance, driving on segments with higher speed limit, and driving during weekends were the factors that increased the odds of getting involved in multiple vehicle crashes leading to fatality or injury. Posted speed limit was a variable that increased the odds of injury/fatal crashes for both analyses.

The literature has indicated that previous violations can be used to predict future offenses and crashes. Therefore, the current study used violation data, in addition to crash data, to examine the groups that are more likely to violate the laws, which account for the majority of driver actions in single truck and multiple vehicle crashes. The violations leading to truck crashes were identified from more than 800 violations and the risk was examined. The results indicated that nonresidents of Wyoming were at higher risk of driving too fast for conditions and failure to drive within single lane. Driving during off peak hours and on weekends also increased the odds of violating these laws.

About 50% of all the multiple vehicle crashes were caused by non-truck drivers. Therefore, both truck and non-truck violations were included in this analysis. Violations related to dominant causes of multiple vehicle crashes were identified and the violation analysis was run only on those violations. The variables included in this analysis were similar to single truck crashes, with the difference that following too close was also added to the contributory factors. Overall, results indicated that being a truck driver plays a dominant role in these types of violations. Truck drivers were more at risk of violating the laws that lead to about 26% of all the causes of multiple vehicle crashes. Driving off peak hours also increased the odds of being involved in violations such as driving too fast for conditions and failing to keep proper lane.

RECOMMENDATIONS

Soole *et al.* (2013) reported that enforcement was effective in changing driver behavior resulting in reduced crash rates [39]. Studies also showed that 90% improvement in law compliance can be achieved by enforcement approaches

[40]. Therefore, WHP can play an important role in reducing the high truck crash rate in Wyoming by allocating resources in the correct areas.

The following recommendations can be given based on this study:

- Nonresidents of Wyoming were more likely to be involved in risky violations that can result in truck related crashes. Therefore, the WHP should educate the nonresidents about the hazards associated with driving in mountainous areas in Wyoming.
- Distraction in the cabin was another variable that increased the odds of getting involved in injury/fatal single truck crashes. Distracted driving laws need to be toughened.
- Weather and lighting conditions increased the odds of being involved in multiple vehicle crashes. The WYDOT should educate the drivers about the risk of driving on not clear weather and night time conditions by displaying speed related messages on dynamic message signs (DMS).
- Truck drivers were more likely to be involved in driving too fast for condition, failing to keep proper lane, and following too close. Stricter measures should be taken for truck drivers who violate these laws.
- Weekend driving increased the odds of injury crashes for both single and multiple truck crashes. Weekend driving also increased the odds of being involved in risky violations. Only 30% of the WHP resources were dedicated to law enforcement on these days. More activity should take place on these days.

It is recommended for the future studies to incorporate the impact of different active safety systems such as adoptive cruise control, forward collision warning and braking, road sign detection, and night vision in their studies.

CONFLICT OF INTEREST

The authors declare no conflict of interest, financial or otherwise.

ACKNOWLEDGEMENTS

The Wyoming Technology Transfer (WYT2/LTAP) center at the University of Wyoming provided extensive resources to assist in the compilation of the data sets used. WYDOT provided the funding of this study. Special acknowledgement goes to WHP and WYDOT for providing the data used in this research.

REFERENCES

1. American Trucking Association. Trucking revenues top $700 billion for the first time according to new report 2015
2. Margaret Weber M, Weber WL. Productivity and efficiency in the trucking industry: accounting for traffic fatalities. Int J Phys Distrib Logist Manag 2004; 34(1): 39-61.
3. Margaret Weber M, Weber WL. Productivity and efficiency in the trucking industry: Accounting for traffic fatalities. Int J Phys Distrib Logist Manag 2004; 34(1): 39-61.
4. Fagin MD. Speed Enforcement. Insurance institute for highway safety 2014.
5. Margaret Weber M, Weber WL. Productivity and efficiency in the trucking industry: Accounting for traffic fatalities. Int J Phys Distrib Logist Manag 2004; 34(1): 39-61.
6. Mashhadi MM, Wulff SS, Ksaibati K. Utilizing crash and violation data to assess unsafe driving actions. J of Sustainable Development of Transport and Logistics 2017; 2(2): 35-46.
7. Rezapour Mashhadi MM, Saha P, Ksaibati K. Impact of traffic Enforcement on Traffic Safety. Int J Police Sci Manag 2017. 1461355717730836.
8. Khattak AJ, Schneider RJ, Targa F. Risk factors in large truck rollovers and injury severity: Analysis of single-vehicle collisions. Transp Res Rec 2003; 40.
9. Chen F, Chen S. Injury severities of truck drivers in single- and multi-vehicle accidents on rural highways. Accid Anal Prev 2011; 43(5): 1677-88.
10. Martensen H, Dupont E. Comparing single vehicle and multivehicle fatal road crashes: A joint analysis of road conditions, time variables and driver characteristics. Accid Anal Prev 2013; 60: 466-71.
11. Craft R. The Large Truck Crash Causation Study Analysis Brief: LTCCS Summary 2007.
12. Lemp JD, Kockelman KM, Unnikrishnan A. Analysis of large truck crash severity using heteroskedastic ordered probit models. Accid Anal Prev 2011; 43(1): 370-80.
13. Moomen M, Rezapour Mashhadi MM, Ksaibati K. An investigation of influential factors of downgrade truck crashes: A logistic regression

approach. J Traff Transp Eng 2018; 1(3): 12-22.

14. Schneider IV, Savolainen P, Zimmerman K. Driver injury severity resulting from single-vehicle crashes along horizontal curves on rural two-lane highways. Transp Res Rec 2009; (2102): 85-92.
15. Zhu X, Srinivasan S. A comprehensive analysis of factors influencing the injury severity of large-truck crashes. Accid Anal Prev 2011; 43(1): 49-57.
16. Pahukula J, Hernandez S, Unnikrishnan A. A time of day analysis of crashes involving large trucks in urban areas. Accid Anal Prev 2015; 75: 155-63.
17. Zou W, Wang X, Zhang D. Truck crash severity in New York city: An investigation of the spatial and the time of day effects. Accid Anal Prev 2017; 99(Pt A): 249-61.
18. Li G, Baker SP. Prior crash and violation records of pilots in commuter and air taxi crashes: A case-control study. Aviat Space Environ Med 1994; 65(11): 979-85.
19. Elliott MR, Waller PF, Raghunathan TE, Shope JT, Little RJ. Persistence of violation and crash behavior over time. J Safety Res 2001; 31(4): 229-42.
20. Chen W, Cooper P, Pinili M. Driver accident risk in relation to the penalty point system in British Columbia. J Safety Res 1995; 26(1): 9-18.
21. Lantz B, Loftus J. Development and implementation of driver safety history indicator into roadside inspection selection system. Transp Res Rec 1969; (2006): 121-5.
22. Terrill TT, Mashhadi MM, Ksaibati K. Developing a tool to help highway patrol in allocating resources to crashes. Int J Police Sci Manag 2016; 1461355716665855.
23. Abdel-Aty M, Uddin N, Pande A, Abdalla F, Hsia L. Predicting freeway crashes from loop detector data by matched case-control logistic regression. Transp Res Rec 1897; (2004): 88-95.
24. Al-Ghamdi AS. Using logistic regression to estimate the influence of accident factors on accident severity. Accid Anal Prev 2002; 34(6): 729-41.
25. Dissanayake S, Lu J. Analysis of severity of young driver crashes: Sequential binary logistic regression modeling. Transp Res Rec 1784; (2002): 108-14.

26. Shinstine DS, Wulff SS, Ksaibati K. Factors associated with crash severity on rural roadways in Wyoming. J of traffic and transportation engineering (English edition) 2016; 3(4): 308-23.
27. Moomen M, Rezapour Mashhadi MM, Ksaibati K. An Investigation of Influential Factors of Truck Crashes on Two-Lane Downgrades in Wyoming: A Logistic Regression Approach. The 97th Transportation Research Board Annual Meeting 2018; 18 05285
28. Kutner MH, Nachtsheim C, Neter J. Nachtsheim, and J. Neter, Applied linear regression models., McGraw-Hill/Irwin, pp. 1-1415, 2004
29. SAS 94 Output Delivery System: User's Guide 2014.
30. Kim JK, Ulfarsson GF, Kim S, Shankar VN. Driver-injury severity in single-vehicle crashes in California: A mixed logit analysis of heterogeneity due to age and gender. Accid Anal Prev 2013; 50: 1073-81.
31. Kim JK, Ulfarsson GF, Kim S, Shankar VN. Driver-injury severity in single-vehicle crashes in California: A mixed logit analysis of heterogeneity due to age and gender. Accid Anal Prev 2013; 50: 1073-81.
32. Abdel-Aty MA, Radwan AE. Modeling traffic accident occurrence and involvement. Accid Anal Prev 2000; 32(5): 633-42.
33. Krull K, Khattak A, Council F. Injury effects of rollovers and events sequence in single-vehicle crashes. Transp Res Rec 1717; (2000): 46-54.
34. Bunn TL, Slavova S, Struttmann TW, Browning SR. Sleepiness/fatigue and distraction/inattention as factors for fatal versus nonfatal commercial motor vehicle driver injuries. Accid Anal Prev 2005; 37(5): 862-9.
35. O'Donnell CJ, Connor DH. Predicting the severity of motor vehicle accident injuries using models of ordered multiple choice. Accid Anal Prev 1996; 28(6): 739-53.
36. Aarts L, van Schagen I. Driving speed and the risk of road crashes: A review. Accid Anal Prev 2006; 38(2): 215-24.
37. Lynam D, Hummel T. The effect of speed on road deaths and injuries: Literature review The Administration; 2002
38. Dobson A, Brown W, Ball J, Powers J, McFadden M. Women drivers' behaviour, socio-demographic characteristics and accidents. Accid Anal Prev 1999; 31(5): 525-35.

39. Soole DW, Watson BC, Fleiter JJ. Effects of average speed enforcement on speed compliance and crashes: A review of the literature. Accid Anal Prev 2013; 54: 46-56.

40. Gains A, Nordstrom M, Heydecker B, Shrewsbury J. The national safety camera programme: Four-year evaluation report 2005

CHAPTER

6

Predicting Truck At-Fault Crashes Using Crash and Traffic Offence Data

Mahdi Rezapour[1], Shaun S. Wulff[2], Khaled Ksaibati[1]

[1] Department of Civil & Architectural Engineering University of Wyoming Office: EN 3084 1000 E University Ave, Dept. 3295 Laramie, WY 82071, USA

[2] Department of Statistics University of Wyoming Office: Ross Hall 336, USA

ABSTRACT

Introduction

The number of truck-related injuries and deaths can be reduced by understanding the factors that contribute to the higher risk of truck-related

Citation: Mahdi Rezapour, Shaun S. Wulff, Khaled Ksaibati, "Predicting Truck At-Fault Crashes Using Crash and Traffic Offence Data", the Open Transportation Journal, vol. 12, 2018, pp. 128, DOI: 10.2174/18744478018120100128.

crashes and violations. Truck drivers are at fault of more than 80% of all the truck crashes on Wyoming interstates, and the literature review indicated that in order to identify appropriate countermeasure to crashes, each crash type should be analyzed individually. The literature review also revealed that relationships exist between driving records and driver culpability in crashes.

METHOD

This study employed two approaches to identify contributory factors to truck-at-fault fatal and injury crashes, and truck-related violations. Interstate 80, a Wyoming corridor in a mountainous area with one of the highest truck crash rates in the US, was selected as a case study. Only truck-at-fault crashes and specific types of truck-related violations were considered in this study. The analyses include two approaches. First, the logistic regression model was employed to explore vehicle, driver, crash, and environmental characteristics that contribute to truck-at-fault fatal and injury crashes. Second, truck violations were used as a proxy for truck crashes to examine the tendency to violate truck-related traffic laws in relation to driver and temporal characteristics. Based on the literature, only violations associated with higher risk of severe crashes were included in the analyses. The included violations accounted for more than 70% of all the violations.

RESULT

This study considered more than 30 variables and found that only 10 variables impact truck-at-fault crashes. These factors included: gender, history of past violation, crashes involving multiple vehicles, exceeding the speed limit, occupant distraction, driver ejection, fatigued driving, non-seat belt usage, overturn, and head-on collision. Results of the second analysis indicated that both residency and time of crash are factors that impact truck-related violations. Results of the analysis also indicated that both residency and time of the crash are factors that impact truck-related violations.

Keywords: Truck-at-fault crash, Severe crashes, Injury, Fatality, Logistic regression, Enforcement, Citation, Risky violation.

INTRODUCTION

A variety of businesses depend on the trucking industry to deliver products in a timely fashion. In the US, the trucking industry delivers more than 80 percent of all the transported freight which accounts for 10 billion tons of commodities consumed with $700.4 billion worth of goods [1].

The economic cost of the crashes in the United States during 2010 alone totaled $871 billion, which includes productivity loss, medical costs, insurance costs, Emergency Service Costs (EMS), workplace losses, and quality of life valuation [2]. The rate of truck-involved crashes has been consistently dropping since 2001. However, the involvement rate of trucks in crashes, is still higher than no-truck vehicles crashes, at a rate of 1.34 per Million Vehicle Mile Traveled (MVMT) compared with 1.28 MVMT [3]. Wyoming has one of the highest percentages of large truck involvement in fatal crashes in the US (23.6%), while the national average is 8.7% [4]. This high rate has resulted mainly from high truck traffic through I-80 and the mountainous areas of Wyoming [5]. According to the US Department of Transportation (DOT), truck drivers caused 87% of truck-related crashes, while only 13% of these crashes were due to other factors such as vehicle and adverse road/weather conditions [6]. Different reasons are attributed to the drivers, including recognition errors, decision errors and performance errors [7].

Traffic crashes can be prevented by legislation or enforcement which control speeding, alcohol consumption, and traffic safety mandates [8]. The countermeasures can be organized into the 4 E's of injury prevention: engineering (*e.g.* highway design), enforcement *(e.g.* local law enforcement agencies), education (*e.g.* advocacy groups), and emergency (*e.g.* fire rescue) [9].

This paper adopted two approaches to reduce truck-at-fault crashes by identification of the contributory factors to injury and fatal (injury/fatal) truck crashes on Interstate 80 (I-80). I-80 is a transcontinental highway in the US that runs from San Francisco in California to Teaneck in New Jersey. The Wyoming section of this interstate is chosen as a case study as this section has one of the highest truck crash rates in the US. The two approaches utilized different explanatory variables including driver, vehicle, environmental, temporal, roadway, and crash factors that are likely to affect injuries and fatalities in truck-related crashes and truck violations on I-80. First, factors contributing to truck crashes, when the truck driver was at fault, were considered. Second, groups that are more prone to the offenses which are more likely to lead to severe truck crashes were identified. The findings of the second analyses will help to evaluate the risk of different truck driver groups in violating traffic laws. Based on the literature review, these violations are associated with an increase of being involved in severe crashes.

LITERATURE REVIEW

A considerable amount of literature has been published on the identification of factors contributing to severe truck crashes. Dong *et al.* identified some of the factors contributing to the severity of truck-involved crashes [10]. Traffic characteristics, driver behavior, and environmental conditions were included in this study. The results revealed that traffic, segment length, the degree of horizontal curvature, and terrain type were some of the factors with effects on car-truck crashes. Woodrooffe and Blower studied truck driver injuries related to crashworthiness. Rollovers, ejection, and frontal impact were identified as the crash types associated with the most severe driver injuries [11]. Lemp *et al.* used the Large Crash Causation Study (LTCCS) to investigate the impact of occupant, driver, vehicle, and environmental characteristics on injury severity in heavy-duty trucks [12]. The results indicated that the number of trailers, truck length, and gross vehicle weight impact truck crash severity.

Naik *et al.* investigated the effects of weather condition on single-vehicle truck crash injury severity [13]. They found that rain, humidity, wind speed, and temperature have significant impacts on truck crash severity. Neeley and Richardson examined the effect of truck-specific regulations on fatality rate by considering truck-involved crashes [14]. They discovered that lower speed limit and truck-length limitations resulted in a reduction of truck-related fatalities. Zhu and Srinivasan evaluated factors affecting large-truck injury crashes [15]. The results indicated that truck driver distraction, alcohol use, and emotional factors of car drivers can contribute to higher severity crashes. Islam investigated contributory factors to injury severity for truck-at-fault crashes [16]. The results indicated that there are different contributory factors to single-vehicle and multiple-vehicle in large truck-at-fault crashes. The literature review also indicated that there are substantial differences between crash types, and therefore each crash type should be analyzed separately [5, 17].

It is believed that a significant percentage of crashes result from driver violation and errors [18]. Therefore, it is important to identify truck driver behavior in violating the traffic laws with the goal of developing effective crash countermeasures. Brace *et al.* examined the relationship between road safety, off-road criminal activity, and traffic offenses of the drivers [19]. They found a positive relationship between criminal activity, traffic offenses, and the likelihood of these drivers being involved in a fatal or serious injury crash. Ayuso *et al.* evaluated the impact of traffic violations on the associated cost of traffic crashes [20]. It was found that some traffic violations were associated with a higher risk of being involved in a serious or fatal crash. They also found out that the costliest traffic violations were speeding-related violations. Associations between traffic offenses and involvement in severe crashes were found in the previous studies [21]. Murray identified violations that were associated with an increase in truck crash likelihood [22]. The factors included risky driving, hours-of-service violation, speeding, and past crash records. Studies also included violation data, in addition to crash data, to improve traffic safety in a more efficient way [5, 17]. Khattak *et al.* evaluated risk factors for truck injury severity [23-25]. The results indicated that risk factors for single-vehicle truck crashes include: risky driving habits of a truck driver such as speeding, reckless driving, and Driving Under the Influence (DUI) as well as roadway geometry and trucks that haul hazardous materials.

Most studies investigated the effects of different variables on the probability of severe crashes. However, far too little attention has been paid to the effects of those variables on the probability of truck-at-fault crash severity. Moreover, no study included violation data, as a proxy safety indicator for crashes, to identify the groups that are at higher risk of involvement in injury/fatal truck crashes. Thus, the current study has two objectives:

- (Crash Analysis) Determine the factors impacting injury/fatal truck crashes when trucks are at fault.
- (Violation Analysis) Predict future severe truck-at-fault crashes using risky truck traffic offenses.

The findings of this second objective will help in identifying the risks of different truck driver groups in violating particular traffic laws. These traffic laws are shown to be some of those associated with increase in the probability of being involved in severe crashes.

DATA PREPARATION

Crash data was obtained from the Wyoming Department of Transportation (WYDOT) using the Critical Analysis Reporting Environment (CARE) package from 2011 through 2014. This study used crash, driver, and Commercial Motor Vehicle (CMV) data. The crash data contains information on reported crashes such as the location, crash circumstances, involved vehicles, and vehicle maneuvering. Driver data includes information on driver attributes such as demographic characteristics, residency of the CMV driver, drug use, and mental condition at the time of a crash. The commercial motor vehicle data contains information related to the CMV itself and driver information such as safety equipment use, driver distraction, and CMV driver violation history. The road geometry characteristics were also obtained from WYDOT. At the scene of the crash, driver condition and crash severity were designated by the police officer according to prescribed categories. Crashes were screened to only include the crashes for at-fault trucks, which weighed at least 10,000 pounds. The resulting dataset for this rural setting had too few fatal crashes for valid statistical modeling. Thus, injury and fatal crashes were included as one category (injury/fatal). The other category is Property Damage Only (PDO), which does not result in any type of injury or fatality.

The citation data was originally obtained from the Wyoming Courts reported violation database from the years 2011- 2014. The data was filtered from the original file to include only truck-related violations. Initially, there were about 85,000 violations on the corridor, I-80, filtered to about 10,661 truck-related violations. About 1000 truck-related violation types were categorized into 7 groups including speeding, seat belt, DUI, Crash Predictor (CP), Hours-Of-Service (HOS), and vehicle-related violations. If a violation did not belong to any of these seven categories, it was grouped under the other violation category. The HOS regulation has been imposed on drivers of commercial vehicles violating maximum driving limits, 14 consecutive hours, or daily off-duty requirements. Violating HOS could result in fatigued driving. Crash Predictor violation includes mostly risky or reckless driving which can result in a crash [26]. DUI violation is related to the drivers who operate a vehicle under the influence of alcohol or drugs.

METHODS

Logistic regression is commonly used in many research studies involving binary crash outcomes [27-29]. Logistic regression models were used to identify factors contributing to crash severity for truck-at-fault crashes. The response (Y) had two possible values: the value 1 if a truck-at-fault crash resulted in an injury or fatality and the value 0 if a truck-at-fault crash resulted in no injury.

Logistic regression was also used for the violation analyses. The purpose of these analyses were to identify truck drivers who were more at risk of committing particular traffic law violations, which may increase the probability of truck crashes that could result in injury/fatality. In this case, the response (Y) had two possible values: the value 1 if a truck driver received a citation of a particular type and the value 0 if a truck driver did not receive a citation of a particular type. The types of citations examined in this study included speeding, DUI, risky driving (Crash Predictor), fatigued driving or Hours Of Service (HOS), and failure to use a seatbelt. Based on the literature review presented previously, speeding, DUI, risky driving (crash predictors), HOS violations and failure to use a seat belt are likely to be associated with increased probability of a crash being severe. Vehicle-related citations were also included in the analysis due to the high percentage of citations in this category (25% of the total number of citations). The violations included in the analyses accounted for 73% of all violations. The probability that a truck driver received a citation is examined in relation to driver characteristics such as age, gender, and residency, as well as temporal characteristics such as time of day and day of week. As described in Kutner *et al.*, the response Y is assumed to have a Bernoulli distribution with probability π [30]. For the crash analysis, this is the probability that a truck-at-fault crash resulted in a fatality or injury. For the citation analysis, this is the probability that a truck driver receives a citation of a particular type. The probability was allowed to depend upon a collection of n predictors ($\boldsymbol{x}$) of interest using the logit link in which

$$\pi(\boldsymbol{x}) = \frac{e^{g(x)}}{1+e^{g(x)}} \text{ or } g(\boldsymbol{x}) = \ln\frac{\pi(x)}{1-\pi(x)} = \beta_0 + \beta_1 x_1 + \beta_2 x_2 + \beta_3 x_3 + \ldots + \beta_n x_n$$

To develop a final logistic regression model, stepwise model selection was used with all possible explanatory variables. Wald based Chi-square tests were conducted using the logistic procedure in SAS. A significance level of 0.10 was specified for entering the model and a significance level of 0.05 was specified for staying in the model. Effects remaining in the model were interpreted using the estimated odds ratio and 95% confidence intervals for the true odds ratio. A standard test for goodness-of-fit in logistic regression is the Hosmer-Lemeshow test [30, 31]. This test was used to assess the fit of the logistic regression model. The overall prediction ability of the model was assessed using the area under the Receiver Operating Characteristic (ROC) curve [32].

VARIABLE DESCRIPTION

Table (**1**) presents descriptive statistics of truck-at-fault crashes and truck-related violations. Since there were more than 30 variables examined in this study, summary statistics are only provided for the significant variables in Table (**1**). Of the 2,042 truck-at-fault crashes, there were 1,695 no injury (PDO) (83%) and 352 were injury/fatal crashes, which include 150 non-incapacitating crashes (7%), 122 unknown and possible crashes (6%), 61 incapacitating crashes (3%), and 19 fatal crashes (<1%). About 95% of truck-at-fault crashes involved male drivers and just 5% involved female drivers. Only 100 (5%) drivers at fault were residents of Wyoming. About 65% of truck-at-fault crashes involved a single vehicle. Over-turn crashes accounted for about 21% of all the truck crashes and front-side crashes comprised one-third of all the truck crashes. Less than 1% of the truck-at-fault crashes involved drivers under the influence and 4% of the drivers showed some sign of fatigue at the time of crashes. Most of the crashes (66%) occurred at higher speed limits (>65) compared with lower speed limit zones (<65). Only 2% of the truck drivers had more than 1 traffic violation on their record before the crashes. Just 41 (2%) of the truck drivers did not use seat belts at the time of the crash. In addition, 551 (27%) of the truck drivers had some types of distraction such as TV, cell pager, or wireless communication inside the cabin at the time of crashes.

For the violation data, the majority of the violations were speeding (26%), vehicle-related (25%), and other violations (26%). About 97% of the drivers violating the traffic laws were male, and the majority of the violators (93%) were non-residents of Wyoming.

RESULTS

Factors Associated with Higher Risk of Truck-at-Fault Injury/Fatal Crashes, Crash Data

The significance of the predictors of crash severity was examined using binary logistic regression. Table (**2**) presents the estimated odds ratios and 95% confidence intervals for the true odds ratio of those predictors that had p-values less than 0.05. The predictors are categorized according to various characteristics and driver behaviors. The models were developed based upon 2,042 truck-at-fault crashes along the I-80 corridor in Wyoming. However, there were missing values in the crash record for some of the predictors. The reduced model was based upon 1,643 truck-at-fault crashes.

Table 1. Summary statistics of the significant explanatory variables.

Truck crashes				
Variable name			**Number**	**%**
Gender Female		Male	1,898	95
		101	5	
Residency Non resident		Resident of Wyoming	102	5
		1,940	95	
Crash characteristics Multiple vehicles involved in a crash Frontal crash Overturn crash		Single vehicle involved in a crash	1,327	65
		715	35	
		668	33	
		425	21	
Driver ejection		CMV Driver is ejected at the crash scene	48	2.35
Driver behavior Fatigued		suspicion of DUI	10	1
		84	4	
Posted speed limit Greater than 65		Less than 65	691	34
		1,342	66	
Violation record CMV driver received more than 1 violation		CMV driver received 1 or less ticket in the past	1,652	98
		34	2	
Safety equipment in use Some types of safety such shoulder or lap belt used		No safety equipment used	41	2
		2,001	98	
Distraction at the time of the crash CMV driver had no distraction at the time of the crash		CMV driver had some distraction at the time of the crash	551	27
		1,491	73	
Crash severity	PDO	PDO	1,695	83
	Injury /fatality	Non-incapacitating injury	150	7
		Unknown/possible injury	122	6
		Incapacitating injury	61	3
		Fatal crashes	19	1
		Truck violation		
Violation Type Vehicle-related Crash predictor HOS violation Seatbelt DUI Others		Speeding	2,800	26
		2,711	25	
		1,454	14	
		385	4	
		277	3	
		123	1	
		2,815	27	

Truck crashes			
Variable name		Number	%
Driver demographic Female	Male	10,335	97
	326	3	
Residency Non resident	Resident of Wyoming	783	7
	9,898	93	

As can be seen in Table (**2**), a few driver and roadway characteristics were identified as significant predictors. Female truck drivers had higher odds of injury/fatal crashes by an estimated 261 percent compared with male truck drivers. O'Donnell and Connor suggested that females are generally less able to sustain different levels and types of physical trauma [33]. Females are not more likely to be involved in a crash, but when they are involved in a crash, it is more likely to be injury/fatality [17]. Truck drivers who had more than one violation are estimated to be 48 percent more likely to be involved in injury/fatal crashes than truck drivers with at most one violation. This result is consistent with previous studies, which indicated drivers with more previous traffic offenses are more likely to be involved in a crash [34, 35]. Results also showed that when the speed limit increased from less than 65 to greater than 65 miles per hour, the estimated odds of an injury/fatal crash increased by 48 percent. The present finding is in contrast with Malyshkina and Mannering, which indicated that speed limit has no effect on crash severity on interstate highways in Indiana [36].

Table 2. Estimated Odds ratio for at fault truck collision with other vehicles, 2011-2014.

Collision Characteristic	**Predictors**	**Odds ratio**	**95 % CL**
Driver Characteristics	**Gender:** Male truck driver (0 if true; 1 otherwise)	2.61*	1.54-4.44
	Age group[1]**:** Younger driver: age <45 years (1 if true; 0 otherwise)		
	Violation record: Traffic record of one ticket or less (0 if true; 1 otherwise)	1.48*	1.10-1.99
	Residency: State of Wyoming (0 if true; 1 otherwise)	-	-
Environmental Characteristic	**Weather condition**[1]**:** Clear (0 if true; 1 otherwise)	-	-
	Road condition[1]**:** Dry (0 if true; 1 otherwise)	-	-
Vehicle characteristic	**CMW weight:** Truck greater than 26000 pounds (1 if true; 0 otherwise)	-	-

Collision Characteristic	Predictors	Odds ratio	95 % CL
Temporal characteristic	**Day of week[1]:** Weekends (1 if true; 0 otherwise)		
	Time[1]: Peak hours 6< <22: (0 if true; 1otherwise)	-	-
Roadway characteristic	Delta[1]	-	-
	Radios[1]	-	-
	Horizontal Length[1]	-	-
	Sag A[1]: Less than 2.0 (0 if true; 1 otherwise)	-	-
	Crest A[1]: Greater than -2 (0 if true; 1 otherwise)	-	-
	Posted speed limit: Speed limit less than 65 (0 if true; 1 otherwise)	1.48*	1.07-2.03
Crash characteristic	**No of vehicle:** 1 (0 if true; 1 otherwise)	3.15*	2.37-4.18
	Pre-collision vehicle actions[1]: Straight-ahead (0 if true; 1 otherwise)	-	-
	Manner of collision		
	Rear-end[1] (1 if true; 0 otherwise)	-	-
	Sidewipe[1] (1 if true; 0 otherwise)	-	-
	Others[1]	-	-
	Head-on (1 if true; 0 otherwise)	1.83*	1.35-2.46
	Rollover (1 if true; 0 otherwise)	4.35*	3.08-6.13
	Occupant ejection: Driver is partially or totally ejected (1 if true; 0 otherwise)	4.55*	1.62-12.83
Driver behavior	**Driver distraction:** No distraction in truck (0 if true; 1 otherwise)	2.57*	1.76-3.75
	DUI suspicion: Driver was suspected of driving under the influence condition (1 if true; 0 otherwise)	-	-
	Sign of fatigue: Driver was fatigued (1 if true; 0 otherwise)	3.69*	2.281-5.990
	CMV driver safety technology in use: Safety equipment was used (1 if true; 0 otherwise)	9.64*	3.03-30.65

Severity (response): no injury=0, injury or fatality=1
*indicates $p<0.05$
[1] indicates insignificant variable included in the initial model

A number of significant crash characteristics were identified by the analyses. Crashes involving multiple vehicles were estimated to be more than three times more likely to result in an injury or fatality compared with single-vehicle crashes. This finding was in agreement with previous research by Khorashadi *et al.*, in which large truck crashes were investigated [37]. Head-on collision increased the odds of injury/fatal crashes by an estimated 83 percent compared with other types of crashes. Rollover crashes led to increased odds of injury/fatal crashes by an estimated 435 percent compared with crashes with other point of impacts.

This finding supports the study by Rezapour *et al.*, who analyzed contributory factors of truck crashes [17]. In their study, higher odds of severe crashes in rollover crashes were linked to an increased contact of a driver with road and vehicle. The odds of involvement in an injury/fatality when the driver was ejected from the vehicle was estimated to be about 455 percent higher than when the driver was not ejected.

A number of important driver behaviors were identified as important predictors of injury/fatal truck-at-fault crashes. Distraction is defined by the National Highway Traffic Safety Administration (NHTSA) as any distraction, including in-vehicle technologies, which takes the eyes or mind of a driver off the road [38]. Distraction was identified as the most common cause of traffic crashes [39]. Driver distraction caused by in-vehicle technologies increased the odds of injury/fatal crashes by an estimated 257 percent compared to that of the non-distracted truck drivers, confirming the previous study conducted on truck crashes [17]. The results also indicated that being fatigued at the time of crash increased the odds of injury/fatal crashes by an estimated 369 percent. This finding was consistent with the study conducted by McCartt *et al.*, who indicated that fatigue was one of the significant factors behind serious truck-involved crashes [40]. Not having any type of CMV safety in use at the time of crashes increased the odds of injury/fatal crashes by an estimated amount of more than 964 percent. This finding confirmed research carried out by Chen *et al.* showing that truck drivers without any use of seat belts are at increased risk of injury and death [41].

The factors presented in Table (**2**) can be classified into three categories based on the estimated odds ratios. However, the uncertainty should also be noted in these comparisons.

As can be seen from Table (**3**), the factor with the highest impact on injury/fatal crashes included the usage of driver safety technology. Number of vehicles, roll-over, occupant ejections, and sign of fatigue were factors that affected truck injury/fatal crashes moderately. Lower impact was observed for factors such as gender, violation record, posted speed limit, head-on collision and driver distraction.

Table 3. Significant contributory factors to truck crash severity.

Degree of Impact	Factors
High (odds ratio estimate >9)	CMV driver safety in use
Medium (3< odd ratio estimate <5)	No of Vehicle
	Rollover
	Occupant ejection
	Sign of fatigue
Low (odd ratio estimate <3)	Gender
	violation record
	Posted speed limit
	Head-on collision
	Driver distraction

The Hosmer-Lemeshow test did not an indicate lack-of-fit for this reduced model of truck-at-fault crash severity (Chi-square = 6.82, df = 9, p-value = 0.5561). The area under the ROC curve for this reduced model was 0.75, which indicates acceptable prediction ability (38).

Factors Associated with Higher Odds of Truck Severe Crashes, Violation Data

This section was set forward to identify the groups that are more likely to be involved in risky violations and, consequently, severe crashes. These violations accounted for more than 70% of all violations. The results of the speeding violation analysis indicated that being a non-resident of Wyoming increased the odds of being involved in a speeding violation by an estimated 68 percent. Two predictors were significant for vehicle-related violations. Wyoming residents were more often exposed to this violation than out-of-state drivers. Drivers were also less likely to violate vehicle-related regulation during off-peak hours, possibly due to lack of visibility by patrols. For crash predictor violations, three predictors were significant. Non-resident drivers were estimated to be more than two times more likely to receive these types of citations. In addition, these violations had an estimated 35% higher odds of occurring during off-peak hours and an estimated 18% higher odds of occurring during weekends. Only peak hours was significant for modeling DUI violations. Three predictors were significant for HOS violations. Both non-residents and off-peak drivers had a higher probability of violating HOS regulation, with estimates of 302% and 78%, respectively. Driving at night decreased the odds of violating the seat belt law by about an estimated 22%.

The Hosmer-Lemeshow test did not provide an indication of lack-of-fit for any of these full models as all p-values exceed 0.25. Therefore, the models provide an indication of the driver and temporal characteristics pertaining to these types of violations. (Table **4**)

Table 4. Adjusted odds ratio for tuck related violation, 2011-2014.

Groups		Speed-ing	Vehicle-related	Crash Pre-dictor	DUI	HOS	Seat Belt
Driver charac-teristics	**Driver gender** Male truck driver (0 if true; 1 otherwise)	-	-	-	-	-	-
	Age of driver: <45(0 if true; 1 otherwise)	-	-	-	-	-	-
	Residence group: Non residence(1 if true; 0 otherwise)	1.68*	0.37*	2.11*		3.02*	-
Tem-poral charac-teristics	**Time:** peak 6< <22(0 if true; 1 otherwise)	-	0.71*	1.35*	1.82*	1.78*	0.67*
	Date: weekends(1 if true; 0 otherwise)	-	-	1.18*	-	0.78	-

*indicates p<0.05 - indicates insignificant variables

DISCUSSION

Crash data from WYDOT was used to investigate the effects of different variables on truck-at-fault injury/fatal crashes. The literature review indicated that there are substantial differences in contributory factors by crash types, so crash types should be analyzed separately. Also, trucks are at fault in more than 80% of all truck- related crashes. Therefore, it is important to identify the underlying causes of these types of crashes as contributory factors to this types of crashes are unique. The literature review was an indication that there are associations between previous violations and future crashes. Therefore, violation data, in addition to crash data, were used in this study to identify the contributory factors to truck-at-fault crashes.

While much research has been done to evaluate factors contributing to truck crash severity, this work focuses on the rural mountainous interstate of I-80 in Wyoming, which has one of the highest truck crash rates in the country. Moreover, the majority of previous studies used all types of truck crashes while this study just included truck-at-fault crashes. In addition, no study used violation data as a proxy safety measure for crashes to identify groups at higher risk of involvement in truck-at-fault crashes. Truck-at-fault crashes were analyzed in this study instead of all types of truck crashes due to the following considerations. First, the literature review indicated that there are substantial differences between contributory factors across different crash types. Second, trucks are at fault in more than 80% of truck-related crashes on Wyoming interstates.The results of crash data analysis showed that a variety of factors affect the odds of a truck-at-fault injury/fatal crashes. In particular, these predictors included violation record, posted speed limit, occupant ejection, driver distraction, fatigued driving, and the use of safety equipment. This current study is unique as it used violation data in addition to crash data. Although any type of violation can increase the odds of crashes, the literature review indicated that some violations are more likely to result in crashes, especially severe crashes. Speeding-related, crash predictor, DUI, HOS, seatbelt violations were included in the analyses. Factors associated with these types of violations could increase the odds of involvement of future truck violations and severe truck crashes. Vehicle-related violations also were included in the analyses due to the high percentage of citations in this category. In addition, only truck-related violations were included in the analyses as the objective was to identify factors associated with truck-at-fault crashes. The included violations accounted for the majority of all the violations (73%), which may justify the included groups of violations.

CONCLUSION

Overall, the results of violation analyses indicated that Residency plays a dominant role in truck-related violations. Non-residents of Wyoming had higher odds of receiving speeding, risky driving and HOS violations, but lower odds of receiving a vehicle-related violation. The odds of receiving the risky truck violations and consequently truck crashes, could be due to a lack of familiarity with driving conditions in Wyoming. However, residents of Wyoming were more likely to violate vehicle-related laws. Time of violation was another important predictor of truck violations. The odds of being involved in risky driving and violating HOS regulations were higher at off-peak hours compared with peak hours. However, night-time truck drivers had smaller odds of receiving a citation for not using a seat belt. The results of the violation analyses can help in understanding tendencies in involvement in risky traffic offenses. The results showed which groups of drivers were more prone to violate the laws that increased the possibility of being involved in severe truck crashes.

RECOMMENDATION

Based on crash and violation analyses, a number of countermeasures could be recommended to Wyoming Highway Patrol (WHP) to reduce the number of injury/fatal truck crashes. These countermeasures included educating truck drivers on the hazards associated with distraction and fatigued driving. A strategy should be employed for truck drivers who have more than one traffic violation due to the fact that these drivers were more susceptible to being involved in injury/fatal crashes. Regarding violation analysis, non-residents of Wyoming are more likely to violate those traffic laws, which could result in injury/fatal truck crashes. This may be due to lack of familiarity of non-residents with the mountainous road conditions of Wyoming highways.

Therefore, it is recommended that WYDOT educate non-residents about the hazards associated with these roads. On the other hand, residents were more likely to violate the laws related to truck weight, size regulations, or issues related to truck vehicles. Stronger enforcement is also recommended between 10 PM and 6 AM to stop HOS violators as these drivers were more likely to violate this law during this time period. Crash predictors or risky driving were also more likely to occur during off-peak hours.

CONFLICT OF INTEREST

The authors declare no conflict of interest, financial or otherwise.

ACKNOWLEDGEMENTS

The Wyoming Technology Transfer (WYT2/LTAP) center at the University of Wyoming provided extensive resources to assist in the compilation of the data sets used. WYDOT provided the funding for this study. Special acknowledgement goes to WHP and WYDOT for providing the data used in this research.

REFERENCES

1. Holcomb R. When trucks stop, America stops 2006.
2. Blincoe L, Miller TR, Zaloshnja E, Lawrence BA. The economic and societal impact of motor vehicle crashes, 2010. 2014.
3. Traffic Safety Facts: 2012 Data 2014. [online]
4. Preliminary statement of policy concerning automated vehicles 2013; 1-14.
5. Mashhadi MMR, Wulff SS, Ksaibati K. Utilizing crash and violation data to assess unsafe driving actions. J Sust Dev Trans Logi 2017; 2(2): 35-46.
6. Department of Transportation, Highway Safety Improvement Program. 2017
7. National Highway Traffic Safety Administration. National motor vehicle crash causation survey: Report to congress. National Highway Traffic Safety Administration Technical Report DOT HS 2008; 811-059.
8. Peden M, Scurfield R, Sleet D, Mohan D, Hyder AA, Jarawan E, et al. World report on road traffic injury prevention. World Health Organization 2004.
9. Federal Highway Admisntartion. Available from: http://safety.fhwa.dot.gov/hsip/resources/fhwasa1102/flyr3_in.cfm 2013.
10. Dong C, Nambisan SS, Richards SH, Ma Z. Assessment of the effects of highway geometric design features on the frequency of truck involved crashes using bivariate regression. Transp Res Part A Policy Pract 2015; 75: 30-41.
11. Woodrooffe J, Blower D. 2015. Heavy Truck Crashworthiness: Injury Mechanisms and Countermeasures to Improve Occupant Safety.
12. Lemp JD, Kockelman KM, Unnikrishnan A. Analysis of large truck crash severity using heteroskedastic ordered probit models. Accid Anal Prev 2011; 43(1): 370-80.
13. Naik B, Tung LW, Zhao S, Khattak AJ. Weather impacts on single-vehicle truck crash injury severity. J Safety Res 2016; 58: 57-65.
14. Neeley GW, Richardson LE Jr. The effect of state regulations on truck-crash fatalities. Am J Public Health 2009; 99(3): 408-15.
15. Zhu X, Srinivasan S. A comprehensive analysis of factors influencing the injury severity of large-truck crashes. Accid Anal Prev 2011; 43(1): 49-57.

16. Islam S, Jones SL, Dye D. Comprehensive analysis of single- and multi-vehicle large truck at-fault crashes on rural and urban roadways in Alabama. Accid Anal Prev 2014; 67: 148-58.
17. Mashhadi MMR, Wulff SS, Ksaibati K. A comprehensive study of single and multiple truck crashes using violation and crash data Open Trans J 2018; 12(1)
18. Rosenbloom T, Perlman A. Tendency to commit traffic violations and presence of passengers in the car. Transp Res, Part F Traffic Psychol Behav 2016; 39: 10-8.
19. Brace C, Whelan M, Clark B, Oxley J. The relationship between crime and road safety 2009.
20. Ayuso M, Guillén M, Alcañiz M. The impact of traffic violations on the estimated cost of traffic accidents with victims. Accid Anal Prev 2010; 42(2): 709-17.
21. Ho KM, Rao S, Burrell M, Weeramanthri TS. The journey from traffic offender to severe road trauma victim: destiny or preventive opportunity? PLoS One 2015; 10(4): e0122652.
22. D Murray, B Lantz, S Keppler. Predicting truck crash involvement: Developing a commercial driver behavior model and requisite enforcement countermeasures. Transportation Research Board 85th Annual Meeting; 2006.
23. Terrill TT, Mashhadi MMR, Ksaibati K. Developing a tool to help highway patrol in allocating resources to crashes. Int J Police Sci Manag 2016; 18(4): 231-41.
24. Rezapour Mashhadi MM, Saha P, Ksaibati K. Impact of traffic Enforcement on Traffic Safety. Int J Police Sci Management 2017; 1461355717730836.
25. Rodríguez D, Rocha M, Khattak A, Belzer M. Effects of truck driver wages and working conditions on highway safety: Case study. Transp Res Rec 1833; (2003): 95-102.
26. Weber A, Murray DC. Evaluating the impact of commercial motor vehicle enforcement disparities on carrier safety performance. Amer Trans Res Inst 2014.
27. Shinstine DS, Wulff SS, Ksaibati K. Factors associated with crash severity on rural roadways in Wyoming. J traf trans engi (English edition) 2016; 3(4): 308-23.
28. Moomen M, Rezapour Mashhadi MM, Ksaibati K. An investigation

of influential factors of truck crashes on two-lane downgrades in wyoming: A logistic regression approach. Transportation Research Board 97th Annual Meeting 2018.

29. Al-Ghamdi AS. Using logistic regression to estimate the influence of accident factors on accident severity. Accid Anal Prev 2002; 34(6): 729-41.
30. Kutner MH, Nachtsheim C, Neter J. Applied linear regression models 2004.
31. Hosmer DW Jr, Lemeshow S, Sturdivant RX. Appl logi reg 2013.
32. Hosmer DW Jr, Lemeshow S, Sturdivant RX. Appl logi reg 2013.
33. O'Donnell CJ, Connor DH. Predicting the severity of motor vehicle accident injuries using models of ordered multiple choice. Accid Anal Prev 1996; 28(6): 739-53.
34. Banks WW, Shaffer JW, Masemore WC, Fisher RS, Schmidt CW Jr, Zlotowitz HI. The relationship between previous driving record and driver culpability in fatal, multiple-vehicle collisions. Accid Anal Prev 1977; 9(1): 9-13.
35. Lui K, Marchbanks P. A study of the time between previous traffic infractions and fatal automobile crashes, 1984–1986. J Safety Res 1990; 21(2): 45-51.
36. Malyshkina N, Mannering F. Effect of increases in speed limits on severities of injuries in accidents. Transp Res Rec 2008; (2083): 122-7.
37. Khorashadi A, Niemeier D, Shankar V, Mannering F. Differences in rural and urban driver-injury severities in accidents involving large-trucks: An exploratory analysis. Accid Anal Prev 2005; 37(5): 910-21.
38. Ranney TA, Mazzae E, Garrott R, Goodman MJ. NHTSA driver distraction research: Past, present, and future. Driver distraction internet forum; 2000.
39. Wang J, Knipling RR, Blincoe LJ. The dimensions of motor vehicle crash risk. J Transp Stat 1999; 2(1): 19-43.
40. McCartt AT, Hellinga LA, Solomon MG. Work schedules of long-distance truck drivers before and after 2004 hours-of-service rule change. Traffic Inj Prev 2008; 9(3): 201-10.
41. Chen GX, Collins JW, Sieber WK, et al. Vital signs: seat belt use among long-haul truck drivers-United States, 2010. MMWR Morb Mortal Wkly Rep 2015; 64(8): 217-21.

CHAPTER 7

Road Traffic Crash Data: An Overview on Sources, Problems, and Collection Methods

Azad Abdulhafedh

University of Missouri-Columbia, MO, USA

ABSTRACT

Road traffic crash data are useful tools to support the development, implementation, and assessment of highway safety programs that tend to reduce road traffic crashes. Collecting road traffic crash data aims at gaining a better understanding of road traffic operational problems, locating hazardous road sections, identifying risk factors, developing accurate diagnosis and remedial measures, and evaluating the effectiveness of road safety programs. Furthermore, they can be used by many agencies and

Citation: Abdulhafedh, A. (2017), "Road Traffic Crash Data: An Overview on Sources, Problems, and Collection Methods". *Journal of Transportation Technologies*, **7**, 206-219. doi: 10.4236/jtts.2017.72015.

businesses such as: law enforcements to identify persons at fault in road traffic crashes; insurers seeking facts about traffic crash claims; road safety researchers to access traffic crash reliable database; decision makers to develop long-term, statewide strategic plans for traffic and highway safety; and highway safety administrators to help educate the public. Given the practical importance of vehicle crash data, this paper presents an overview of the sources, trends and problems associated with road traffic crash data.

Keywords: Road Safety, Vehicle Crash Data, Over-Dispersion, Under-Dispersion, Under-Reporting, FARS, NASS, HSIS

INTRODUCTION

Throughout the world, cars, buses, trucks, motorcycles, pedestrians, animals, taxis and other categories of travelers, share the roadways, contributing to economic and social development in many countries. Yet each year, many vehicles are involved in crashes that are responsible for millions of deaths and injuries. Globally, every year, about 1.25 million people are killed in motor vehicle crashes and approximately 50 million more are injured. Vehicular crashes are the world's leading cause of death for individuals between the ages of one and twenty-nine [1] . Following current trends, about two million people could be expected to be killed in motor vehicle crashes each year by 2030 [1] . Currently, road crashes are ranked as the ninth most serious cause of death in the world, and without new initiatives to improve road safety, fatal crashes will likely rise to the third place by the year 2020 [1] . In developed countries, road traffic death rates have decreased since the 1960s because of successful interventions such as seat belt safety laws, enforcement of speed limits, warnings about the dangers of mixing alcohol consumption with driving, and safer design and use of roads and vehicles. For example, road traffic fatalities have declined by about 25.0 percent in the United States from 2005 to 2014 and the number of people injured has decreased 13.0 percent from 2005 to 2014 [2] . In Canada, the number of road traffic fatalities has declined by about 62.0 percent from 1990 to 2014, and the number of injuries has declined by about 68.0 percent during the same period [3] . However, traffic fatalities have increased in developing countries from 1990 to 2014 (i.e. 44.0 percent in Malaysia and about 243.0 percent in China) [1] . Developing countries bear a large share of the burden, accounting for 85.0 percent of annual deaths and 90.0 percent of the disability-adjusted life years. More than one-half of all road traffic deaths globally involve people ages 15 to 44, during their most productive earning years. Moreover, the disability burden for this age group accounts for about 60.0 percent of all disability-adjusted life years. The costs and consequences of these losses are significant. Three-quarters of all poor families who lost a member in a traffic crash reported a decrease in their standard of living, and about 61.0 percent reported having to borrow money to cover expenses following their loss [4] . The World Bank estimates that road traffic injuries cost 2.0 percent to 3.0 percent of the Gross National Product of developing countries, or twice the total amount of development aid received worldwide by developing countries [5] . Crash-related fatalities and injuries can be prevented or at least minimized by a joint involvement from multiple sectors (i.e. transportation agencies, police, health departments, education institutions) that oversee road safety, vehicles, and the drivers themselves.

Effective interventions include design of safer infrastructure and incorporation of road safety features into land-use and transport planning; improvement of vehicle safety features; improvement of post-crash care for victims of road crashes, and improvement of driver behavior, such as setting and enforcing laws relating to key risk factors, and raising public awareness [6] . In addition, vehicular crash data can assist with the development of generalized theories concerning road safety.

A range of basic laws have been put forth to help explain the relationship between the occurrence of road crashes and potential risk factors, such as: the universal law of learning, which implies that the crash rate tends to decline as the number of kilometers travelled increases; the law of rare events, which states that rare events, such as environmental hazards, would have more effect on crash rates than regular events; and the law of complexity, which implies that the more complex the traffic situation road users encounter, the higher the probability of crash occurrence [7] .

Although transportation agencies often try to identify the most dangerous road sites, and put great efforts into preventive measures, such as illumination and policy enforcement, the annual number of traffic crashes has not yet significantly decreased. For instance, 35,092 traffic fatalities were recorded in the US during 2015, an increase of 7.2% as compared to the previous year [8] . The fatality rate per 100 million vehicle miles traveled increased 3.7% from 2014-2015.

Thirty-five States had more motor vehicle fatalities in 2015 than in 2014. Given this trend, it is imperative to gain a better understanding of crash data sources, trends and problems.

THE IMPORTANCE OF COLLECTING VEHICULAR CRASH DATA

Vehicular crash data are used to respond to requests from the congress, federal agencies, state and local governments, universities and research organizations, highway safety communities, the media, and private citizens. Accurate data are required to support the development, implementation, and assessment of highway safety programs aimed at reducing crash tolls. An example of the practical importance of collecting and maintaining vehicular crash data is the recent emerging of the crash data retrieval tools, commonly referred to as the vehicle black boxes. Based upon a rule imposed by the National Highway Traffic Safety Administration (NHTSA), most vehicles manufactured and sold in North America after 2012 are equipped with Event Data Recorders (EDRs) that collect, store, and retrieve vehicle crash event data. The EDRs can help law enforcement investigating vehicle crashes to recover crucial crash data parameters from a vehicle that has been involved in a crash, including pre-crash data that will help better understand important factors that led to the crash occurrence [9] . Another practical example is the use of the Crash Outcome Data Evaluation System (CODES), which is a program managed by NHTSA, to link crash records to injury outcome records collected at the scene by emergency medical services. CODES data has been utilized to improve traffic safety issues in different ways, such as examining whether the increased crash rates for teen drivers have resulted in an increased injury to their passengers, and exploring the seat belt usage in preventing injuries and fatalities. CODES data has also been used to inform and educate traffic safety decision-makers at federal, state, and local levels in many circumstances, for instance, providing federal and state legislators with CODES reports on the importance of seat belt use in preventing injuries and fatalities; delivering data to the state highway administrations to develop long- term, statewide strategic plans for traffic and highway safety; and publishing CODES fact sheets that can help educate the public [10] .

ROAD TRAFFIC DATA COLLECTION METHODS

Most studies of traffic related problems begin with the collection of data. Generally, traffic data collection methods can be classified as one of two categories: intrusive and non-intrusive methods. Intrusive methods typically involve a data recorder and a sensor placing on or in the road [11] . The most common intrusive devices are:

- Pneumatic road tubes: rubber tubes placed across the road lanes to detect vehicles from pressure changes that are produced when a vehicle tire passes over the tube. The pulse of air that is created is recorded and processed by a counter located on the side of the road. The main drawback of this technology is that it has limited lane coverage and its efficiency is subject to weather, temperature and traffic conditions.
- Piezoelectric sensors: sensors are placed in a groove along roadway surface of the lane(s) monitored. The principle is to convert mechanical energy into electrical energy. The amplitude and frequency of the signal is directly proportional to the degree of deformation.
- Magnetic loops: this is the most conventional technology used to collect traffic data. The loops are embedded in roadways in a square formation that generates a magnetic field. The information is then transmitted to a counting device placed on the side of the road. This has a generally short life expectancy because it can be damaged by heavy vehicles, but is not affected by bad weather conditions.

Non-intrusive techniques are based on remote observations ranging from human observation to those based on new technologies [12] :

- Manual counts: Trained observers gather traffic data such as vehicle occupancy rate, pedestrians and vehicle classifications that cannot be efficiently obtained through automated counts. Equipment needs are rather basic with the observers usually requiring only a tally sheet, mechanical and/or electronic counting devices.
- Passive and active infra-red sensors: the presence, speed and type of vehicles can be detected based on the infrared energy radiating from the detection area. The main drawbacks of this method are the sensor's performance during bad weather, and limited lane coverage.
- Passive magnetic sensors: magnetic sensors can be fixed under or on top of the roadbed. The sensors record the number of vehicles, their type and speed. However, in some operating conditions, the sensors have difficulty differentiating between closely spaced vehicles.

- Microwave radar sensors: these sensors can detect moving vehicles and record vehicle counts, speed and vehicle classification and are not usually compromised by weather conditions.
- Ultrasonic and passive acoustic sensors: these devices emit sound waves to detect vehicles by measuring the time for the signal to return to the device. The ultrasonic sensors can be placed directly over the lane or alongside the road to collect vehicle counts, speed and classification data However, the collection ability of these sensors can be adversely affected by temperature or bad weather.
- Video image detection: video cameras can be used to record vehicle numbers, type and speed by means of different video techniques e.g. trip line and tracking. Video detection systems can be sensitive to weather conditions.

The Floating Car Data (FCD) can be used to collect traffic data by locating the vehicle via mobile phones or GPS over the entire road network. Data such as car location, speed and direction of travel can then be sent anonymously to a central processing center. After being collected and extracted, useful information can be redistributed to the drivers on the road [13] .

There are two important traffic measures that are widely used in modeling traffic data, namely: the average annual daily traffic (AADT); and the vehicle miles travelled (VMT). These two traffic variables, usually derived from fixed sensors measurements, play a key role in traffic crash analysis and policy decisions [14] . AADT is the average (calculated over a year) number of vehicles passing a point along a particular counting section each day. Thus, AADT represents the vehicle flow over a road section (e.g. highway segment) on an average day of the year. Methods for calculating AADT are generally based on data from two types of counts: permanent automatic traffic counts and short-period traffic counts. A combination of these two measurements is generally used to obtain an AADT estimate over a larger road network. In the US, the factoring method is a common methodology used to estimate AADT. This method has been adopted by many transportation agencies as a standard protocol corresponding with federal guidelines. The 2013 Traffic Monitoring Guide serves as a reference document that provides general guidance on the development of traffic monitoring programs for highway agencies. In particular, the TMG provides guidance on the collection of traffic volume, vehicle classification, and weight information [15] . VMT refers to the distance travelled by vehicles.

It is often used as an indicator of traffic demand and for analyzing mobility patterns and travel trends. It plays a key role in various important decision-makings such as air quality compliance, roadway pavement maintenance, and crash analysis. There are four methods commonly used to calculate VMT [16] :

- Odometer readings (vehicle-based method) at regular vehicle inspections, the average distance travelled by the vehicles is determined and then multiplied by the number of road vehicles.
- Traffic counts (road-based method) for one considered link, the VMT is calculated by multiplying the AADT by the length of the link. VMT for a roadway can then be obtained by summing the VMT of each segment.
- Driver survey questionnaires sent to households with one or more cars soliciting information such as the number of miles driven by each vehicle during the whole year and unit consumption.
- Fuel consumption the volume of road traffic is estimated from information about fuel supply and fuel consumption as derived from estimates of miles driven per fuel gallon for typical types of vehicles.

SOURCES OF VEHICULAR CRASH DATA

In the U.S., a variety of efforts to collect, maintain and/or distribute information on vehicular crash data have been utilized. Some of the crash data sources that are publicly available are listed below:

Fatality Analysis Reporting System (FARS)

FARS is an online database of fatal motor vehicle crashes that documents all fatalities that occurred within the 50 States since 1975. FARS qualifying crashes had to involve a motor vehicle traveling on a public traffic way, and must have resulted in the death of a motorist or a non-motorist within 30 days of the crash. FARS is administered by the National Center for Statistics and Analysis (NCSA) within the National Highway Traffic Safety Administration (NHTSA). FARS data are collected from each State's government by trained state employees, who are responsible for gathering, and transmitting their state's data to NCSA in a standard format. After the data file is created, quality checks are performed on the data, and the electronic data are made available online to the public in Statistical Analysis System (SAS) data files as well as Database Files (DBF).The main SAS data files include: the Accident file, which contains information about crash characteristics and environmental conditions at the time of the crash; the Vehicle file, which contains information describing the in-transport motor vehicles and the drivers of in-transport motor vehicle who are involved in the crash; the Person file, which contains information describing all persons involved in the crash including motorists and non-motorists (e.g., pedestrians); the Damage file, which contains information about all areas on the vehicle that were damaged in the crash; the Drimpair file, which contains information about physical impairments of drivers of motor vehicles; the Factor file, which contains information about vehicle circumstances that may have contributed to the crash; the Violatn file, which contains information about violations that were charged to drivers; and the Vindecode file, which contains vehicle descriptors based on the vehicle's VIN. The temporal coverage of FARS data includes some variables such as, the time of the crash, the date, the month, and the year. The spatial coverage of FARS data includes the latitude and longitude coordinates of each crash location. The FARS data are generally complete, reliable, and publicly available online [17] . However, one of the FARS data weaknesses is that FARS data cannot be downloaded for multiple years at a time due to the system complexities, and when data is downloaded from FARS website, the user can obtain data by only one variable at a time. In addition, as mentioned above, the FARS data does not provide the injury-severity only crashes, and property-damage only crashes.

The NASS-GES

The National Automotive Sampling System (NASS)-General Estimates System (GES) obtains its data from a representative crash sample selected from more than five million police-reported crashes annually in the US. These crashes include those that result in a fatality or injury and those involving major property damage as well. The data are obtained by NASS-GES data collectors in 60 geographic sites across the United States. These data collectors make visits to approximately 400 police agencies within the 60 sites, where they randomly sample about 50,000 crashes per year. NASS-GES data are made available to the public in Statistical Analysis System (SAS) data files as well as Database Files (DBF). The main SAS data files of NASS-GES include similar FARS files mentioned above. The temporal coverage of the NASS-GES data includes variables such as, time of the crash, the date, the month, and the year. The spatial coverage only includes the land use of the crash location without providing the latitude and longitude of the crash location or the x, y coordinates. One weakness in NASS- GES data is that it uses a weighted data element that produces the overall national estimates that may differ from the true state-level values because they are based on a probability sample of crashes among the country, and this cannot give the accurate state-level estimates, which decreases the reliability of the data. Another weakness is that the NASS-GES data are obtained either directly from the police accident report (PAR) or by interpreting the information provided in the PAR through reviewing the crash diagram, or combinations of data elements on the PAR. Because of this interpretation, an important portion of data can be missing in the system [18] .

The NASS-CDS

The National Automotive Sampling System (NASS)-Crashworthiness Data System (CDS) obtains its data from 24 geographic sites in the US. These data are weighted to represent all police reported motor vehicle crashes occurring in the USA during the year including light vehicles, such as, passenger cars, SUVs, and vans. The NASS-CDS files are available in a Statistical Analysis System (SAS) dataset, and contain similar FARS files. The NASS-CDE system provides temporal coverage of data through variables such as, time of the crash, the date, the month, and the year. There is no spatial coverage within the NASS-CDS data, as it does not provide the latitude and longitude of the crash location nor the x, y coordinates. One weakness of the NASS-CDS data is that the data from these crashes are weighted to produce national estimates, and cannot give the state- level estimates, which decreases the reliability of data [19] .

The State Data System (SDS)

The State Data System (SDS) is maintained by NHTSA's National Center for Statistics and Analysis (NCSA), and only thirty-two states are participating in the system, including the state of Missouri. While the (FARS) only has fatal crash data, SDS provides data on injury and property-damage-only crashes as well. In contrast to the data in (NASS-GES), the SDS consists of census data taken directly from police accident reports. The law enforcement agencies within a state are the primary source of information on crashes occurring within a state. All states have requirements for documenting fatal, injury or property damage crashes (with damage above a certain dollar threshold). Each participating state has its own reporting system, for instance, in the state of Missouri, the Missouri Statewide Traffic Accident

Records System (STARS) is managed by the Missouri State Highway Patrol (MSHP), and all Missouri law enforcement agencies are required by law to submit a Missouri Uniform Traffic Crash Report to STARS if a traffic crash occurred that involves a death, a personal injury, or a property damage. STARS involves many recording files, such as, the Crash and Personal Severity, which includes fatal, personal injury, and property damage; the Crash Circumstances file, which includes motorcycles crashes by year; Speed Involved Traffic Crash file; Alcohol Involved Traffic Crash file; Young Driver Involved Traffic Crash file; and Mature Driver Involved Traffic Crash file. All files are provided in excel and PDF format, complete, reliable, and available online for the public (MSHP 2016). The temporal coverage of the SDS data includes variables such as, time of the crash, the date, the month, and the year. The spatial coverage only includes the x, y coordinates of the crash locations in only some spots. One weakness of the SDS data is that it does not provide a comprehensive list of risk variables and details that exist in the FARS and NASS-GES systems [20] .

The Highway Safety Information System (HSIS)

The Highway Safety Information System (HSIS) is a highway data system funded by the U.S. Federal Highway Administration (FHWA), with data voluntarily provided to HSIS by the participating states, which are California, Washington, Minnesota, Illinois, Ohio, Maine, and North Carolina. HSIS began operation in 1987, and the participating states were selected based on their data availability, quantity, and quality of data. HSIS supports the FHWA safety research program, and can be accessed online by researchers, universities, and safety professionals.

The HSIS files are available in a (SAS) format, and the main files include four basic files namely; the Accident file, the Vehicle file, the Occupant file, and the Roadway file.

The temporal coverage of the HSIS data includes variables such as, time of the crash, date, month, and the year. The spatial coverage only includes the section length, and the milepost of the crash location without providing the latitude and longitude of the crash location nor the x, y coordinates. The HSIS data are generally complete with very few missing data, reliable, and publicly available. One weakness of the HSIS data is that it does not cover all states within the US, and also their main files should be merged in order to get the required information [21] .

Data.Gov

The Data.gov is a federal open US government online database that includes all states, and local government's metadata describing their open data resources. Data.gov began operation in 2009, and is managed and hosted by the U.S. General Services Administration, Office of Citizen Services and Innovative Technologies, and follows the Project Open Data schema that includes fields, such as title, description, tags, publisher, etc. for every data set displayed on the website. Different data topics are available, such as Agriculture, Health, Business, Climate, Energy, Finance, and Science. The transportation statistics series consists of analyzed statistical information on motor fuel, vehicle crashes, motor vehicle registrations, driver licenses, highway user taxation, highway mileage, travel, and highway finance. The files are available in CSV format, and can be freely downloaded without registration [22] .

The U.S. Census Bureau

The U.S. Census Bureau is part of the Department of Commerce, and is overseen by the Economics and Statistics Administration. The transportation section within the online database provides data on civil air transportation, water trans- portation, revenues, passenger and freight traffic volume, trains, highway mileage and finances, highway crash data, characteristics of public transit, and railroads. Data are available in excel format for public use [23] .

The SHRP2-NDS

The Strategic Highway Research Program 2-Naturalistic Driving Study (SHRP2- NDS) is an online database related to the Transportation Research Board (TRB)'s second safety project for an in-vehicle driving behavior field study collected from naturalistic driving data and associated participant, vehicle, and crash-related data. The project was conducted by six site contractors located at geographically distributed data collection sites throughout the United States and more than 3000 individuals participated in the study. Given that the SHRP 2-NDS is a federally funded study that involves human subjects, the collection of the data and its use in analysis are subject to the approval of institutional review boards. The SHRP 2-NDS database is managed by the Virginia Tech Transportation Institute, and researchers interested in accessing the data must demonstrate that they are qualified researchers seeking the data for research purposes [24] .

The Center for Advanced Public Safety (CAPS)

The Center for Advanced Public Safety (CAPS) is a research center at the University of Alabama that deals with vehicular crash data, and traffic safety improvements, among other research areas. CAPS has developed a tool for crash data analysis called the Critical Analysis Reporting Environment (CARE), which has many useful analytical functions such as, frequency distributions, cross-ta- bulations, and statistical significance tests. CARE can compare the performance of one subset of data against another in terms of all potential variables that could demonstrate performance differentials. CARE analysis software is free to download and is required to analyze and visualize the electronic data contained within CAPS datasets. The CAPS online crash datasets are free to download, and contains a variety of crash data files that mainly belongs to the state of Alabama, such as the vehicle crash files, the driver data file, the person data file, and the road data file [25] .

COUNT DATA

When discussing traffic crash data, it is important to differentiate between a count, and count data. The term count typically refers to an enumeration of events. Count data, on the other hand, refers to the observations made about events that are enumerated [26] . A common quality of count data is that (0.0) is the most frequently observed value, (1.0) is the next most observed, (2.0) the next, and so on. Use of count data is widespread in many disciplines, including transportation engineering. Examples of count data applications in transportation include the number of driver route changes per day, the number of trip departure changes per week, number of vehicles waiting in a queue, and the number of crashes observed on road segments per some time period, such as a year, or five years. Count data are often described as random events, sporadic (i.e. isolated or scattered), rare, discrete, not continuous, and non-negative integers [27] . One frequent pitfall is to model count data as continuous data by applying an ordinary least square regression [28] . This approach is inappropriate because regression models can produce predicted values that are non-integers and can also predict values that are negative, both of which are inconsistent with count data. In addition, many distributions of count data are positively skewed with many observations in the data set having a value of 0.0. The high number of zeros in the data set prevents the transformation of a skewed distribution into a normal one, which is a requirement of normal distribution. An alternative is to use a Poisson distribution or one of its variants. Poisson distributions have a number of advantages over an ordinary normal distribution, including a skew, discrete distribution, and the restriction of predicted values to non-negative numbers [28] .

COMMON PROBLEMS WITH CRASH DATA

Crash data suffer from some problems or issues that have been identified over the years. These problems are a potential source of error in modeling crash data that may cause incorrect estimates and inferences. These issues are summarized below:

- Over-dispersion: over-dispersion occurs when the observed variance exceeds the theoretical variance of the crash counts, which violates the assumption of the most common count-data modeling approach. Over-dispersion in crash data can result from a variety of factors, such as the clustering of data, unaccounted temporal correlation, and model miss-specification (Cameron and Trivedi 1998). When data are over-dispersed, estimation of a crash model can lead to biased parameter estimates, which in turn could lead to incorrect inferences regarding the factors that determine crash-frequencies [29] [30] [31] [32] .
- Under-dispersion: under-dispersion occurs when the observed variance of the crash counts is smaller than the assumed (i.e. theoretical) variance, and most likely to occur with small sample sizes. Although rare, however, under-dispersion can lead to incorrect parameter estimates and crash prediction [29] [31] [32] [33] .
- Small Sample Size: crash data collection process may be expensive, therefore crash data are sometimes characterized by a small number of observations (i.e. small sample size), which can produce low sample-mean. Small samples can cause estimation problems in traditional count prediction models. For example, with small sample sizes, the maximum likelihood estimation of parameters could produce insufficient results [34] [35] . Also, it was shown that the dispersion parameter of the negative binomial model can be incorrectly estimated when using data characterized by a small sample size and low sample mean [36] .
- Time Interval Variations: crash data are typically collected over some time period, such as one year, three years, and five years.

Over the collection period, some explanatory variables and their relationship to the crash incidents may change a reality that is not usually considered due to the lack of detailed data within the collection period. Ignoring within-period variation in explanatory variables may result in biased estimation of parameters and incorrect prediction of crashes as a result of unobserved heterogeneity [28] [35] .

- Temporal and Spatial Autocorrelations: the prediction of crash models can be improved when several years of crash data are utilized, such as a period of three years instead of one year [37] . However, this means that the same road- way entity will generate multiple observations, which will be correlated over time because many of the unobserved effects associated with a specific roadway entity will remain the same over time. This phenomenon is termed temporal autocorrelation, which can adversely affect the precision of parameter estimates. Similarly, correlation of observations over space can exist given that roadway entities may be in close proximity and may share unobserved effects. This phenomenon is termed spatial autocorrelation and if not appropriately addressed, can also lead to incorrect parameter estimates [38] [39] [40] [41] [42] .
- Omitted-Variables Bias: modeling crash prediction with few explanatory variables could produce simplified models with omitted-variables bias. Leaving out important explanatory variables can result in biased parameter estimates and incorrect inferences, especially if the omitted variable is correlated with variables included in the model, which is often the case [43] [44] [45] [46] .
- Under-Reporting: traffic crash data may suffer from under-reporting effects, especially for minor and less severe crashes. The unknown parameters in the models are generally estimated assuming random sampling from the population, therefore, if under-reporting is not accounted for, then it could result in biased samples that are likely to produce incorrect parameters in the model-estimation process [28] [44] [47] .
- Non-Linear Relationships Bias: many crash prediction models assume that explanatory variables influence the dependent variables in linear manner. However, it has been shown that non-

linear functions can often better characterize the relationships between crash frequencies and explanatory variables. For example, using traffic flow as a measure of exposure, some have found that the crash prediction per unit of exposure becomes smaller as traffic flow increases pointing to unobserved heterogeneity and possible other specification problems in the functional form of the model [39] [48] .

CONCLUSION

Road traffic crash data are useful tools to support highway safety programs that tend to reduce road traffic crashes. They can be used by many authorities such as: law enforcements to identify persons at fault in road traffic crashes; insurers seeking facts about traffic crash claims; road safety researchers to access crash reliable database; decision makers to develop long-term, statewide strategic plans for traffic and highway safety; and highway safety administrators to help educate the public. Given such trends, this paper presented a general overview of the sources, collection methods, and problems associated with crash data to better gaining an understanding of road traffic operational problems, locating hazardous road sections, identifying risk factors, developing accurate diagnosis and remedial measures, and evaluating the effectiveness of road safety programs.

REFERENCES

1. WHO (2015) WHO Global Report on Road Safety 2015. http://www.who.int/violence_injury_prevention/road_safety_status/2015/en/
2. NCSA (2015) NHTSA-National Center for Statistics and Analysis. http://www.nhtsa.gov/NCSA
3. Transport Canada (2016) Road Safety in Canada. http://www.tc.gc.ca/eng/motorvehiclesafety/tp-tp15145-1201.htm
4. Beirness, D.J. and Beasley, E. (2011) A Comparison of Drug- and Alcohol-Involved Motor Vehicle Driver Fatalities. Canadian Centre on Substance Abuse, Ottawa.
5. World Bank (2015) The World Bank—Transport for Development. http://blogs.worldbank.org/transport/why-vehicle-safety-matters-crash-related-deat hs?cid=EXT_WBBlogSocialShare_D_EXT
6. Mohan, D. (2002) Road Safety in Less-Motorized Environments: Future Concerns. International Journal of Epidemiology, 31, 527-532. https://doi.org/10.1093/ije/31.3.527
7. Elvik, R. (2006) Laws of Accident Causation. Accident Analysis and Prevention, 38, 742-747.
8. NCSA (2016) NHTSA-National Center for Statistics and Analysis. http://www.nhtsa.gov/NCSA
9. NHTSA-Ruling (2010) The Crash Data Services. http://www.crashdataservices.net/NHTSAruling.html
10. NHTSA-CODES (2011) The Crash Data Outcome Evaluation System. http://www.nrd.nhtsa.dot.gov/cats/listpublications.aspx?Id=219&ShowBy=Category
11. Bar-Gera, H. (2007) Evaluation of a Cellular Phone-Based System for Measurements of Traffic Speeds and Travel Times: A Case Study from Israel. Transportation Research Part C, 15, 380-391.
12. Fraser, S. (2007) The Use of Floating Cellular Telephone Data for Real-Time Transportation Incident Management. McMaster University, Hamilton.
13. Robichaud, K. and Gordon, M. (2003) Assessment of Data-Collection Techniques for Highway Agencies. Transportation Research Record, 1855, 129-135. https://doi.org/10.3141/1855-16
14. Sliupas, T. (2006) Annual Average Daily Traffic Forecasting Using Different Techniques. Transport, 21, 38-43.

15. Ehlert, A., Bell, M.G.H. and Grosso, S. (2006) The Optimization of Traffic Count Locations in Road Networks. Transportation Research Part B, 40, 460-479.
16. Fricker, J. and Kumapley, R. (2002) Updating Procedures to Estimate and Forecast Vehicle-Miles Traveled. Purdue University, West Lafayette. https://doi.org/10.5703/1288284313337
17. NHTSA-FARS (2016) The Fatality Analysis Reporting System. http://www.nhtsa.gov/FARS
18. NASS-GES (2016) The General Estimate System. https://www.nhtsa.gov/national-automotive-sampling-system-nass/nass-general-estimates-system#11381
19. NASS-CDS (2016) The Crash Worthiness Data System. https://www.nhtsa.gov/national-automotive-sampling-system-nass/crashworthiness-data-system
20. NHTSA-SDS (2016) NHTSA-The State Data System. https://www.nhtsa.gov/state-data-programs/sds-overview
21. HSIS (2016) The Highway Safety Information System. http://www.hsisinfo.org/index.cfm
22. Data.gov. (2016) US Government Data. https://catalog.data.gov/dataset
23. The US Census Bureau (2016) The Data of the US Census Bureau. http://www.census.gov/en.html
24. SHRP2-NDS (2016) The Strategic Highway Research Program 2—Naturalistic Driving Study. https://insight.shrp2nds.us/
25. CAPS (2016) The Center for Advanced Public Safety. http://www.caps.ua.edu/analytics/downloads/datasets/public/
26. Hilbe, J. (2014) Modeling Count Data. Cambridge University Press, Cambridge. https://doi.org/10.1017/cbo9781139236065
27. Hauer, E. (1992) Traffic Conflicts and Exposure. Accident Analysis and Prevention, 14, 359-364.
28. Glenberg, A. (1996) Learning from Data: An Introduction to Statistical Reasoning. 2nd Edition, Lawrence Erlbaum Associates, Mahwah.
29. Cameron, A.C. and Trivedi, P.K. (1998) Regression Analysis of Count Data. Cambridge University Press, Cambridge. https://doi.org/10.1017/CBO9780511814365
30. Miaou, P. (1994) The Relationship between Truck Accidents and

Geometric Design of Road Sections: Poisson versus Negative Binomial Regressions. Accident Analysis and Prevention, 26, 471-482.

31. Park, S. and Lord, D. (2007) Multivariate Poisson-Lognormal Models for Jointly Modeling Crash Frequency by Severity. Transportation Research Record, 2019, 1-6. https://doi.org/10.3141/2019-01
32. Abdulhafedh, A. (2016) Crash Frequency Analysis. Journal of Transportation Technologies, 6,169-180. https://doi.org/10.4236/jtts.2016.64017
33. Oh, J., Washington, S. and Nam, D. (2006) Accident Prediction Model for RailwayHighway Interfaces. Accident Analysis and Prevention, 38, 346-356.
34. Wood, G.R. (2002) Generalized Linear Accident Models and Goodness of Fit Testing. Accident Analysis and Prevention, 34, 417-427.
35. Lord,D.andBonneson,A.(2007)DevelopmentofAccidentModification Factors for Rural Frontage Road Segments in Texas. Transportation Research Record, 2023, 20-27. https://doi.org/10.3141/2023-03
36. Lord, D. (2006) Modeling Motor Vehicle Crashes Using Poisson-Gamma Models: Examining the Effects of Low Sample Mean Values and Small Sample Size on the Estimation of the Fixed Dispersion Parameter. Accident Analysis and Prevention, 46, 751-766.
37. Mohammadi, M., Samaranayake, V. and Bham, G. (2014) Crash Frequency Modeling Using Negative Binomial Models: An Application of Generalized Estimating Equation to Longitudinal Data. Accident Analysis and Prevention, 2, 52-69.
38. Gujarati, D. (1992) Essentials of Econometrics. McGraw-Hill, New York.
39. Lord, D. and Persaud, N. (2000) Accident Prediction Models with and without Trend: Application of the Generalized Estimating Equations Procedure. Transportation Research Record, 1717, 102-108. https://doi.org/10.3141/1717-13
40. Washington, P., Karlaftis, G. and Mannering, F. (2010) Statistical and Econometric Methods for Transportation Data Analysis. 2nd Edition, Chapman Hall, CRC, Boca Raton.
41. Lord, D. and Mannering, F. (2010) The Statistical Analysis of Crash Frequency Data: A Review and Assessment of Methodological Alternatives. Accident Analysis and Prevention, 44, 291-305.
42. Savolainen, P., Mannering, F., Lord, D. and Quddus, M. (2011) The

Statistical Analysis of Highway Crash-Injury Severities: A Review and Assessment of Methodological Alternatives. Accident Analysis and Prevention, 43, 1666-1676.

43. Arminger, G., Clogg, C. and Sobel, M. (1995) Handbook of Statistical Modeling for the Social and Behavioral Sciences. Plenum Press, New York. https://doi.org/10.1007/978-1-4899-1292-3
44. Caliendo, C., Guida, M. and Parisi, A. (2007) A Crash-Prediction Model for Multilane Roads. Accident Analysis and Prevention, 39, 657-670.
45. El-Basyouny, K. and Sayed, T. (2009) Collision Prediction Models Using Multivariate Poisson-Lognormal Regression. Accident Analysis and Prevention, 41, 820- 828.
46. Geedipally, R., Lord, D. and Dhavala, S. (2012) The Negative-Binomial Lindley Generalized Linear Model: Characteristics and Application Using Crash Data. Accident Analysis and Prevention, 45, 258-265.
47. Anastasopoulos, P.C. and Mannering, F. (2009) A Note on Modeling Vehicle Accident Frequencies with Random-Parameters Count Models. Accident Analysis and Prevention, 41, 153-159.
48. Shankar, N., Milton, J.C. and Mannering, F. (1997) Modeling Accident Frequencies as Zero-Altered Probability Process: An Empirical Enquiry. Accident Analysis and Prevention, 29, 829-837.

CHAPTER 8

Road Crash Prediction Models: Different Statistical Modeling Approaches

Azad Abdulhafedh

University of Missouri-Columbia, MO, USA

ABSTRACT

Road crash prediction models are very useful tools in highway safety, given their potential for determining both the crash frequency occurrence and the degree severity of crashes. Crash frequency refers to the prediction of the number of crashes that would occur on a specific road segment or intersection in a time period, while crash severity models generally explore the relationship between crash severity injury and the contributing factors such as driver behavior, vehicle characteristics, roadway geometry, and

Citation: Abdulhafedh, A. (2017), "Road Crash Prediction Models: Different Statistical Modeling Approaches". *Journal of Transportation Technologies*, **7**, 190-205. doi: 10.4236/jtts.2017.72014.

road-environment conditions. Effective interventions to reduce crash toll include design of safer infrastructure and incorporation of road safety features into land-use and transportation planning; improvement of vehicle safety features; improvement of post-crash care for victims of road crashes; and improvement of driver behavior, such as setting and enforcing laws relating to key risk factors, and raising public awareness. Despite the great efforts that transportation agencies put into preventive measures, the annual number of traffic crashes has not yet significantly decreased. For instance, 35,092 traffic fatalities were recorded in the US in 2015, an increase of 7.2% as compared to the previous year. With such a trend, this paper presents an overview of road crash prediction models used by transportation agencies and researchers to gain a better understanding of the techniques used in predicting road accidents and the risk factors that contribute to crash occurrence.

Keywords: Crash Prediction Models, Poisson, Negative Binomial, Zero-Inflated, Logit and Probit, Neural Networks

INTRODUCTION

Road traffic accidents are the world's leading cause of death for individuals between the ages of one and twenty-nine [1] . Throughout the world, cars, buses, trucks, motorcycles, pedestrians, animals, taxis and other categories of travelers, share the roadways, contributing to economic and social development in many countries. Yet each year, many vehicles are involved in crashes that are responsible for millions of deaths and injuries. Globally, every year, about 1.25 million people are killed in motor vehicle crashes and approximately 50 million more are injured. Following current trends, about two million people could be expected to be killed in motor vehicle crashes each year by 2030 [1] . Currently, road crashes are ranked as the ninth most serious cause of death in the world, and without new initiatives to improve road safety, fatal crashes will likely rise to the third place by the year 2020 [1] . In developed countries, road traffic death rates have decreased since the 1960s because of successful interventions such as seat belt safety laws, enforcement of speed limits, warnings about the dangers of mixing alcohol consumption with driving, and safer design and use of roads and vehicles. For example, road traffic fatalities have declined by about 25.0 percent in the United States from 2005 to 2014 and the number of people injured has decreased 13.0 percent from 2005 to 2014 [2] . In Canada, the number of road traffic fatalities has declined by about 62.0 percent from 1990 to 2014, and the number of injuries has declined by about 68.0 percent during the same period [3] . However, traffic fatalities have increased in developing countries from 1990 to 2014 (i.e. 44.0 percent in Malaysia and about 243.0 percent in China) [1] . Developing countries bear a large share of the burden, accounting for 85.0 percent of annual deaths and 90.0 percent of the disability-adjusted life years. More than one-half of all road traffic deaths globally involve people ages 15 to 44, during their most productive earning years. Moreover, the disability burden for this age group accounts for about 60.0 percent of all disability-adjusted life years. The costs and consequences of these losses are significant. Three-quarters of all poor families who lost a member in a traffic crash reported a decrease in their standard of living, and about 61.0 percent reported having to borrow money to cover expenses following their loss [4] . The World Bank estimates that road traffic injuries cost 2.0 percent to 3.0 percent of the Gross National Product of developing countries, or twice the total amount of development aid received worldwide by developing countries [5] . Although transportation agencies often try to identify the most hazardous road sites, and put great efforts into preventive measures, such as illumination and policy enforcement, the annual number of traffic crashes has not yet significantly decreased. For instance, 35,092 traffic fatalities were recorded in the US during 2015, an increase of 7.2% as compared to the previous year [6] . The fatality rate per 100 million vehicle miles traveled (VMT) increased 3.7% between 2014-2015. Thirty-five States had more motor vehicle fatalities in 2015 than in 2014. Every month except November saw increases in fatalities from 2014 to 2015, and the highest increases occurred in July and September [6] . Given this trend, it is imperative to gain a better understanding of the risk factors that may be associated with traffic crashes. This paper aims at presenting an overview of road crash prediction models used by transportation agencies and researchers to help understanding the techniques used in predicting road accidents and the risk factors that contribute to crash occurrence.

THE IMPORTANCE OF TRAFFIC ACCIDENTS PREDICTION MODELS

Traffic accidents prediction models are very useful tools in highway safety, given their potential for determining both the frequency of accident occurrence and the contributing factors that could then be addressed by transportation policies. Vehicular crash data can be used to model both the frequency of crash occurrence and the degree of crash severity. Crash frequency refers to the prediction of the number of crashes that would occur on a specific road segment or intersection in a time period [7] . Crash severity methods generally explore the relationship between crash severity injury categories and contributing factors such as driver behavior, vehicle characteristics, roadway geometry, and road-environ- ment conditions. Traffic accident related-fatalities and injuries can be prevented or at least minimized by a joint involvement from multiple sectors (i.e. transpor- tation agencies, police, health departments, education institutions) that oversee road safety, vehicles, and the drivers themselves. Effective interventions include design of safer infrastructure and incorporation of road safety features into land- use and transport planning; improvement of vehicle safety features; improvement of post-crash care for victims of road crashes, and improvement of driver behavior, such as setting and enforcing laws relating to key risk factors, and raising public awareness [8] . Transportation agencies and research institutions often seek to identify the most dangerous road sites, and this will require modeling road crash data to determine both crash frequency and crash severity degree. In addition, traffic accidents prediction models can also assist with the development of generalized theories concerning road safety. A range of basic laws have been put forth to help explain the relationship between the occurrence of road crashes and potential risk factors, such as: the universal law of learning, which implies that the crash rate tends to decline as the number of kilometers travelled increases; the law of rare events, which states that rare events, such as environmental hazards, would have more effect on crash rates than regular events; and the law of complexity, which implies that the more complex the traffic situation road users encounter, the higher the probability of crash occurrence [9] .

FACTORS AFFECTING ROAD TRAFFIC ACCIDENTS

A traffic accident may have many contributing factors, such as those related to driver behavior, road geometry, traffic volumes, vehicle, and environment. The influence of such variables on crash occurrence could significantly vary on a case-by-case basis, but in general, both behavioral factors related to the driver's errors, and non-behavioral factors related to road geometry, traffic flow conditions, vehicle, and environment are thought to significantly affect traffic crashes [10] . Research has revealed that there are generally six major groups of risk factors affecting traffic crash occurrence [11] [12] [13] [14] [15] :

1) Driver behavior: alcohol and drug use, reckless operation of vehicle, failure to properly use occupant protection devices, the use of cell phones or texting, and fatigue.
2) Vehicle factors: vehicle type, and the engineering and the safety design standards for vehicle performance. For example, the design of windshield glass and the location and durability of gas tanks can increase safety. Passenger protection systems in vehicles (i.e. air bags, safety belts), if used, can eliminate injuries or reduce their severity.
3) Roadway characteristics: road geometries and road side conditions, such as well-designed curves and grades, wide lanes, adequate sight distance, clearly visible striping, flared guardrails, good quality shoulders, roadsides free of obsta- cles, well-located crash attenuation devices, and well-planned use of traffic signals.
4) Traffic volumes: average annual daily traffic (AADT) or the vehicle miles travelled (VMT). AADT is the average number of vehicles passing a point along a particular road section each day. Thus, AADT represents the vehicle flow over a road section on an average day of the year. VMT refers to the distance travelled by vehicles on roads. It is often used as an indicator of traffic demand and is commonly applied to evaluate mobility patterns and travel trends.
5) Environmental factors: weather conditions, and light conditions.
6) Time factors: the season of the year, the month of the year, weekdays, and the hour of crash occurrence.

THE COSTS OF ROAD TRAFFIC ACCIDENTS

The highest cost of traffic crashes is in the loss of human lives; however, society also bears the consequences of many costs associated with motor vehicle crashes. Highway crashes currently cost the USA about $1078.0 billion a year, approximately 5.0 percent higher than 2000. Total costs include both economic costs and societal harm [16] . In the year 2010, 3.9 million people were injured and 32,999 killed in 13.6 million motor vehicle crashes in the US [2] . The economic costs of these crashes totaled $242.0 billion including lost productivity, medical costs, legal and court costs, emergency service costs, insurance administration costs, congestion costs, property damage, and workplace losses. The $242.0 billion cost of motor vehicle crashes represents the equivalent of nearly $784.0 for each person living in the United States, and 1.6 percent of the $14.96 trillion U.S. Gross Domestic Product for 2010 [16] . When quality of life valuation is considered, the total value of societal harm from motor vehicle crashes in 2010 was $836.0 billion, roughly three and a half times the value measured by economic impacts alone. Lost market and household productivity accounted for $77.0 billion of the total $242.0 billion economic costs, while property damage accounted for $76.0 billion. Medical expenses totaled $23.0 billion. Congestion caused by crashes, including travel delay, excess fuel consumption, greenhouse gases and criteria pollutants accounted for $28.0 billion. Each fatality resulted in an average discounted lifetime cost of $1.4 million. Each critically injured survivor cost an average of $1.0 million [16] .

LITERATURE REVIEW

Early crash analysis models were generally based on simple multiple linear regression methods assuming normally distributed errors. However, researchers soon discovered that crash occurrence could be better fitted with a Poisson distribution. Hence, a Poisson regression model based upon a generalized linear framework was soon adopted over conventional multiple linear regression techniques. Several such Poisson regression approaches for exploring the relationship between the risk factors and crash frequency have been proposed [15] [16] [17] [18] [19] [20] . However, it has been found that Poisson regression approaches have one important constraint that the mean must be equal to the variance which if violated, the standard errors estimated by the maximum likelihood method, will be biased, and the test statistics derived from the model will be incorrect. Recent studies have shown that crash data are usually over-dispersed, when the variance exceeds the mean, therefore, incorrect estimation of the likelihood of crash occurrence could result in applications of the Poisson regression model [7] . In efforts to overcome the problem of over-dispersion, researchers began to employ the Negative Binomial (NB) distribution (also called the Poisson-Gamma) instead of the Poisson distribution, which relaxes the mean equals to variance constraint, and hence can accommodate over-dispersion in crash data counts [7] . NB models have been widely used in crash frequency modeling [14] [15] [19] [21] [22] [23] . However, NB models have some limitations such as the inability to handle under-dispersion of crash counts when the mean of the crash counts is higher than the variance. Although rare, this phenomenon can arise when the sample size is very small, leading to erroneous parameter estimates [24] [25] . To address the limitations of NB models, Poisson-lognormal models have been proposed, in which the error term is Poisson-lognormal rather than gamma-distributed to better handle the under-dispersed crash counts [21] [26] [27] . Another widely used type of crash prediction model is the zero-inflated Poisson and zero-inflated negative binomial models, which have been introduced mainly to deal with the over-dispersion problem caused by excessive zeroes (i.e. locations where no crashes can be observed) in traffic data counts. The zero-inflated models have shown great flexibility, although their applicability in crash prediction has been criticized because of the long term mean equals zero in the safe state that could produce some biased estimates [7] [22] . Generalized additive modeling approaches have also been proposed which provide smoothing functions for the explanatory variables. However, these models typically include more parameters than the traditional count models, and therefore their applicability to the crash prediction has been very limited [28] [29] . Random- parameters models have been applied to take the effect of the unobserved heterogeneity from one roadway site to another, however, their application in practice has been very limited [30] [31] [32] . The finding that road crashes are poorly explained by linear functions of independent variables, has encouraged the exploration of non-linear approximators such as fuzzy logic and neural networks. For example, a fuzzy logic approach was used for prediction of urban highway crash occurrence and it was found that the use of fuzzy sets in crash prediction is indeed a viable approach [33] . Neural networks have been applied to highway safety applications as predictive tools, such as in driver behavior analysis, pavement maintenance, vehicle detections, traffic signal control, and vehicle emissions, however, their application to crash analysis has been limited [28] [34] [35] . For instance, an artificial neural network was utilized to analyze the freeway crash frequency in Taiwan, and the results indicated that an artificial neural network can provide a consistent alternative method for analyzing crash frequency [36] . Also, a group of artificial neural networks was applied to model the non-linear relationships between the injury severity levels and crash-related factors. The findings indicated that artificial neural network models can predict crashes more effectively than the traditional statistical methods [37] . In crash severity models, a wide variety of statistical approaches such as the binary and the multinomial logit models, nested logit models, mixed logit models and ordered probit models have been investigated. For example, the ordered probit model was applied to predict crash severity on roadway sections, signalized intersections and toll plazas in Florida [38] . A mixed logit model was applied that used the injury outcome of the crash using limited crash data to investigate the proportion of crashes of each severity level on a specific roadway segment over a specified time period. Then, the number of crashes by severity level was determined without the need for detailed crash-specific data [39] . Also, a multinomial logistic regression was applied to model the severity injury of different vehicle collision patterns in urban highways in Arkansas, and the researchers recommended the use of the MNL over other models [40] .

A REVIEW ON THE STATISTICAL APPROACHES OF ROAD CRASH PREDICTION MODELS

There are different statistical approaches for modeling traffic crashes. The following approaches present some of the mostly used methods.

Multiple Linear Regression

Early models of traffic accident models were based on the simple multiple linear regression approach assuming normally distributed errors. The general form of the linear crash prediction model can be expressed as follows:

$$Y|\theta \sim Dist(\theta) \text{with} \theta = f(X, \beta, \varepsilon) \quad (1)$$

where,

Y: the dependent variable (i.e. crash frequency),

θ: the crash dataset,

Dist(θ): the model distribution,

X: a vector representing different independent variables (i.e. risk factors),

β: a vector of regression coefficients,

f(.): link function that relates X and Y together,

ε: the disturbance or error terms of the model.

Poisson Regression

Although multiple linear regression models have been widely applied, it has been found that crash occurrence can often be better fitted with a Poisson distribution. One frequent pitfall is to model crash data as continuous data by applying an ordinary least square regression [41] . This approach is inappropriate because regression models can produce predicted values that are non-integers and can also predict values that are negative, both of which are inconsistent with continuous data modeling. In addition, many distributions of crash data are positively skewed with many observations in the data set having a value of 0.0. The high number of zeros in the data set prevents the transformation of a skewed distribution into a normal one, which is a requirement of normal distribution. An alternative is to use a Poisson distribution or one of its variants. Poisson distributions have a number of advantages over an ordinary normal distribution, including a skew, discrete distribution, and the restriction of predicted values to non-negative numbers [41] . Hence, generalized linear modeling variates of the Poisson regression model have been proposed to explore the relationship between the risk factors and traffic accident modeling [15] [17] [18] [19] . Poisson regression has been applied to a wide range of transportation count data, including crash frequency. A Poisson regression model is similar to an ordinary linear regression, with two exceptions. First, it assumes that the errors follow a Poisson (not normal) distribution. Second, rather than modeling the response variable Y as a linear function of the regression coefficients, it models the natural log of the response variable, ln(Y), as a linear function of the coefficients [7] . The Poisson model can be expressed as follows:

$$P(n_i) = \frac{\lambda i EXP(-\lambda i)}{n!} \qquad (2)$$

where,

$P(n_i)$: the probability of n crashes occurring on a highway segment i,

n_i: the number of observations per time period (such as a year),

λ_i: the expected crash frequency on road segment i per time period (i.e. the mean of distribution) which can be estimated as follows:

$$\lambda_i = EXP(\beta X_i) \qquad (3)$$

where

X_i: a vector of the independent variables (i.e. risk factors),

β: a vector of the estimates (coefficients) of the independent variables X_i.

This model is estimable by standard maximum likelihood methods, with the log likelihood (LL) function given as:

$$LL(\beta) = \sum_{1}^{n}\left[-EXP(\beta X_i) + n(\beta Xi) - Ln(n!)\right] \quad (4)$$

One assumption of Poisson Models is that the mean and the variance are equal, an assumption that is sometimes violated [7] . This can be dealt with by using a dispersion parameter if the difference is small, or by using a negative binomial regression model if the difference is large [42] .

Negative Binomial Regression Model (NB)

In order to overcoming the problem of over-dispersion, the Negative Binomial (NB) distribution (also called the Poisson-Gamma) has been investigated as an alternative to the Poisson distribution given that it relaxes the condition of mean equals to variance, and hence can take into account over-dispersion in the crash data counts [7] . As a result, NB models have been widely applied in crash frequency modeling [14] [15] [19] [21] [22] [23] .

The NB uses a Gamma probability distribution and can relax the assumption of the mean equals the variance and, hence, the NB can accommodate over-dis- persion that may exist in the crash data counts [43] . A primary source of over- dispersion is the clustering of data, and the possible omission of relevant independent variables influencing the Poisson rate across observations [44] . In order to obtain the NB model, the Poisson regression can be rewritten by adding an error term to its expected number of crashes, and becomes [7] :

$$\lambda i = EXP(\beta Xi + \varepsilon_i) \quad (5)$$

where EXP (ε_i) is a gamma-distributed error with mean equals one and variance equals α. The addition of this term allows the variance VAR (n_i) to differ from the mean E (n_i) as shown in Eq. 6:

$$VAR(n_i) = E(n_i)(1 + \alpha E(n_i)) \quad (6)$$

This error term is called the over-dispersion parameter, and both α and β can be estimated from the maximum likelihood function. When α is zero, the model becomes Poisson regression, and if α is found to be significantly different from zero, then the NB regression can be used instead of the Poisson regression model to handle the over-dispersion in crash data. However, the NB model also has some limitations such as its inability to handle the case of under-dis- persion of the data count, when the mean of the crash counts is higher than the variance [25] [44] .

Poisson-Lognormal Regression Model

To address the limitations of the NB models, the Poisson-lognormal model was introduced, in which the error term is Poisson-lognormal rather than gamma- distributed so as to better handle under-dispersed data counts [21] [26] [27] . The Poisson-lognormal model is similar to the negative binomial model, however, the EXP (ε_i) term used in the model is lognormal-rather than gamma-distributed. The Poisson-lognormal model provides more flexibility than the negative binomial model, but it does have some limitations, such as, its complex estimation of parameters due to the fact that the Poisson-lognormal distribution does not have a closed form [26] .

Zero Inflated Poisson and Negative Binomial Regression Models

Another widely used crash frequency modeling approach is the zero-inflated Poisson and zero-inflated negative binomial models, which have been introduced primarily to deal with the over-dispersion problem caused by excessive zeroes (i.e. locations where no crashes can be observed) in traffic data counts. The zero-altered procedure allows modeling the crash frequencies in two states, namely; the zero-crash state, and the non-zero crash state (where crash frequencies follow Poisson or negative binomial distribution), and the probability of a section being in zero or non-zero states can be found by a binary logit or probit model. In crash data, large numbers of zero observations are commonly present largely due to under reporting of minor crashes at these sites, the presence of dangerous crash sites (i.e. non-zero crash sites) in close proximity to the neighboring zero crash sites rendering the zero-crash sites to the safe mode, and given that some of zero crash sites may be free from only certain type of crashes, not all types of crashes [45] . Zero-inflated models attempt to account for such excess zeros. A dual state crash system may be assumed, in which one state is the zero crash state that can be regarded as virtually safe during the observation period, while the other state is the non-zero crash state. For example, consider vehicle crash occurring per year on 1-kilometer sections of highway. For straight sections of roadway with wide lanes, low traffic volumes, and no roadside objects, the likelihood of a vehicle crash occurring may be extremely small, but still pre- sent because an extreme human error could randomly cause an accident. These sections are considered to be in a zero-crash state that refer to situations where the likelihood of an event occurring is extremely rare in comparison to the non-zero state where crash occurrence is inevitable and follows some count distribution [46] . To address the zero-inflated modeling processes, the zero-inflated Poisson (ZIP) and the zero-inflated negative binomial (ZINB) regression models have been developed. The probabilities of the two possible zero- and non-zero states are: p_i for the zero crash state, and $(1-p_i)$ for the non-zero crash state, and the overall probability of crashes is the sum of the probabilities from each state. The probability of crash frequency in the zero state can be modeled as:

$$Pr(n_i = 0) = p_i + (1 - p_i)R_i(0) \quad (7)$$

where $R_i(0)$ is the probability of zero crashes that occurs in the zero state. The probability of crash frequency in the non-zero state can be modeled as:

$$Pr(n_i > 0) = (1 - p_i)R_i(n_i) \quad (8)$$

where $R_i(n_i)$ is the probability of non-zero crashes in the non-zero state. Maximum likelihood estimates can be used to estimate the parameters of both ZIP and ZINB regression models and confidence intervals are constructed by likelihood ratio tests. In zero-inflated models, the two state process is assumed to follow a logit (logistic) or probit (normal) probability process [45] . Zero-inflated models have shown great flexibility in both states, although their applicability to crash prediction has been criticized because of the long term mean equals to zero in the safe state, and hence, biased estimates may result [7] .

Conway-Maxwell Poisson Regression Models

The Conway-Maxwell Poisson model has been recently investigated with respect to highway safety issues, but its application in crash frequency modeling has been rather limited [7] . Generalized additive models have been explored given that they can provide smoothing functions for the explanatory variables. The Conway-Maxwell Poisson distribution is a generalization of the Poisson distribution that can handle both under-dispersed and over-dispersed crash data. The main advantage of this model is to handle the under-dispersion in crash data that cannot be modeled by the Poisson model or the Negative Binomial model. However, the low sample-mean, and small sample size of the under-dispersed crash data can influence the estimated parameters, and therefore, it has been limited in the application of crash frequency [24] . However, in practice, the estimation of these models can become very difficult as they require more parameters, a problem that has likely impeded their application to crash frequency prediction [29] [47] .

Random-Parameter Models

Random-parameters models have also been investigated to take the effect of the unobserved heterogeneity from one roadway site to another [31] [32] .

The motivation for random-parameter models is to account for unobserved heterogeneity across observations. Random-parameter models can be derived by assuming that the estimated parameters vary across observations according to some distribution. Estimated parameters can be modeled as [48] :

$$\beta_n = \beta + \omega_n \tag{9}$$

where

β_n: a vector of estimated parameters of the n observations,

ω_n: a randomly distributed term.

With this equation, the Poisson, and the Negative Binomial parameters become:

$$\lambda i \middle| \omega_n = EXP(\beta_n X_n) \tag{10}$$

$$\lambda i \middle| \omega_n = EXP(\beta n\ Xn + \varepsilon_n) \tag{11}$$

Artificial Neural Networks and Fuzzy Logic models

Given that a linear function may not sufficiently explain the relationship between the dependent variables and the associated independent variables in crash modeling, non-linear approximators such as fuzzy logic and neural networks have also been explored. Artificial Neural Networks (ANNs) are a class of computational intelligence tools that can be used for prediction and classification problems. ANNs can model very complex non-linear functions to high accuracy levels using a process of learning that is similar to the learning procedure of the cognitive system in the human brain. The network body is composed of input layers, hidden layers, and output layers. These models can be trained to appro- ximate any nonlinear function to a required degree of accuracy using a learning algorithm (such as back propagation) that would give the desired output, in a supervised learning process. ANNs have some advantages over the statistical models. For instance, regression models need a pre-defined relationship or functional form between the dependent variable (crash frequency) and the independent explanatory variables that can be estimated by some statistical approaches, whereas the ANNs do not require the establishment of these functional forms, and can be easily applied in the analysis. On the other hand, the ANNs differ from the statistical models in that they behave as black-boxes and do not provide interpretation for the parameter estimates [15] [18] [35] [36] . Fuzzy logic applications have increasingly been proven to have a significant crash-predicting capability in recent years [49] . Fuzzy logic system is defined as the nonlinear mapping of an input data set to a scalar output data, and the first step of the process (known as fuzzification) consists of gathering a crisp set of input data that will be converted to a fuzzy set using fuzzy linguistic variables, fuzzy linguistic terms, and membership functions. After that, an inference is made based on a set of fuzzy rules, and then, the resulting fuzzy output is mapped to a crisp output using the membership functions, in the defuzzification step [33] .

Logit and Probit Models

Logit and Probit models can be applied to study crash severity modeling. The data used in modeling crash severity is often attributed with many details relating to the crash occurrence (i.e. such as the number of vehicles involved, age of victims, weather conditions, types of vehicles involved, and crash type) which can be integrated in statistical models. Since the dependent variable (i.e. crash severity) usually has two or more outcome categories (i.e. fatal, injury, property-damage-only), logit and probit models are often used to model the severity of crash data. Discriminant analysis could also be used to model crash severity, but given its rigid assumptions, logit and probit models have been viewed as preferable [32] [50] . Binary models consider two response outcomes (i.e. fatal vs. non-fatal or injury vs. property-damage-only), and multinomial models consider three or more response outcomes. Traffic accident severity models can be generally classified as either nominal or ordinal. Although there is no consensus on which model is the best, as the selection of the model is often governed by the characteristics of the data, some researchers have opted for nominal models over ordinal models. The rationale for this choice is likely due to the influence that independent variables in ordinal models could exert on the ordered discrete outcome probabilities. That is, in closely related categories (i.e. no injury and possible injury) there may be some shared unobserved effects among adjacent injury categories. Failing to account for such correlation could generate incorrect inferences [32] [51] . Others still prefer ordinal models due to their simplicity and overall performance, especially when less detailed data are available. Binary models consider two outcomes, and multinomial models consider three or more outcomes. In binomial logit or probit models, the dependent variable, Y, can take one of two values 0.0 or 1.0. For example, injury or non-injury, fatal or non-fatal. The general shape of binomial logit model is (assuming $□_i$ = Pr (Y_i= 1)):

$$\text{Logit}(\pi_i) = \log\left[\frac{\pi_i}{1-\pi_i}\right] == X_i\beta \qquad (12)$$

where,

X_i: a vector of explanatory variables (i.e. risk factors),

β: a vector of regression coefficients.

As □ approaches zero, logit (□) tends toward -∞; and as □ approaches 1.0, logit (□) tends toward +∞ [52] . The binomial probit model is an alternative to the binomial logit model, in which the probit ($□_i$) is the standard cumulative normal distribution function (θ^{-1}) that can be expressed as:

$$\text{Pro}bit(\pi_i) = \theta^{-1}(\pi_i) = Xi\beta \qquad (13)$$

There are many types of the multinomial models that can be used in modeling crash severity, such as, the multinomial logistic regression (MNL), the nested logistic regression, the mixed logistic regression, and the multinomial probit models. For example, The MNL tries to find the best fitted model to describe the relationship between the polytomous dependent variable with more than two categories and a set of independent variables. The logistic regression model is a non-linear transformation of the linear regression model, as it consists of an S-shaped distribution function [53] . The logit distribution constrains the estimated probabilities that lie between 0.0 and 1.0, as shown in Figure 1.0. The logistic regression function is bounded by 0.0 and 1.0, whereas the linear regression function may predict values above 1.0 and below 0.0.The logistic (logit) function can be expressed as:

$$\text{Log}it(p) = b_0 + b_1X_1 + b_2X_2 + \cdots + b_kX_k \qquad (14)$$

where,

p: the probability of presence of an outcome of interest,

X_k: the vector of k independent variables,

b_0: the regression coefficient on the constant term (intercept),

b_k: the vector of regression coefficients on the independent variables X_k,

The odd ratio is the probability of the event divided by the probability of the nonevent, and is defined as follows [50] [53] :

$$\text{odd ratios} = p/(1 - p) \qquad (15)$$

When p = 0, then odd (p) = 0, when p = 0.5, then odd (p) = 1.0, and when p = 1.0, then odd (p) =∞. The logit transformation is defined as the logged odds:

$$\text{Log}it(p) = \ln\left[p/(1-p)\right] \qquad (16)$$

The transformation from odds to log of odds is the log transformation, and this is a monotonic transformation. That is, the greater the odds, the greater the log of odds and vice versa. Logit (p) can be back-transformed to p by the following formula:

$$p = \frac{1}{1+e^{-logit(p)}} \qquad (17)$$

The transformation from probability to odds is a monotonic transformation as well, meaning the odds increase as the probability increases or vice versa. Probability ranges from 0.0 and 1.0. Odds range from 0.0 and positive infinity [53] [54] .

CONCLUSION

Traffic crash prediction models are very useful tools in road safety programs used by transportation agencies, police, health departments, education institutions that oversee road safety, vehicles, and the driver's education. They can be used to predict both the frequency of crash occurrence and the contributing factors that could then be addressed by transportation policies. According to the world health organization (WHO), road crashes are ranked as the ninth most serious cause of death in the world, and present the world's leading cause of death for individuals between the ages of one and twenty-nine. Each year, traffic accidents are responsible for killing about 1.25 million people and injuring approximately 50 million more. Following current trends, about two million people could be expected to be killed in motor vehicle crashes each year by 2030. The World Bank estimates that road traffic injuries cost 2.0 percent to 3.0 percent of the Gross National Product of developing countries. Given a such trend, this paper presented different types of traffic crash prediction models to gain a better understanding of the techniques used to predict road accidents and their contributing risk factors. A wide range of statistical approaches were presented including, Poisson regression, Negative Binomial regression, Zero-Inflated models, logit and probit models, and machine learning methods.

REFERENCES

1. World Health Organization (2015) Global Status Report on Road Safety 2015. http://www.who.int/violence_injury_prevention/road_safety_status/2015/en/
2. NHTSA—National Center for Statistics and Analysis (NCSA) (2016) NHTSA Studies Vehicle Safety and Driving Behavior to Reduce Vehicle Crashes. http://www.nhtsa.gov/NCSA
3. Transport Canada (2016) Road Safety in Canada. http://www.tc.gc.ca/eng/motorvehiclesafety/tp-tp15145-1201.htm
4. Beirness, D.J. and Beasley, E. (2011) A Comparison of Drug- and Alcohol-Involved Motor Vehicle Driver Fatalities. Canadian Centre on Substance Abuse, Ottawa.
5. World Bank (2015) The World Bank-Transport for Development. http://blogs.worldbank.org/transport/why-vehicle-safety-matters-crash-related-deat hs?cid=EXT_WBBlogSocialShare_D_EXT
6. NHTSA—National Center for Statistics and Analysis (NCSA) (2016) NHTSA. https://www.nhtsa.gov/press-releases/us-dot-announces-steep-increase-roadway-de aths-based-2015-early-estimates
7. Lord, D., and Mannering, F. (2010) The Statistical Analysis of Crash Frequency Data: A Review and Assessment of Methodological Alternatives. Accident Analysis and Prevention, 44, 291-305.
8. Mohan, D. (2002) Road Safety in Less-Motorized Environments: Future Concerns. International Journal of Epidemiology, 31, 527-532. https://doi.org/10.1093/ije/31.3.527
9. Elvik, R. (2006) Laws of Accident Causation. Accident Analysis and Prevention, 38, 742-747.
10. Caliendo, C., Guida, M. and Parisi, A. (2007) A Crash-Prediction Model for Multilane Roads. Accident Analysis and Prevention, 39, 657-670.
11. Greibe, P. (2003) Accident Prediction Models for Urban Roads. Accident Analysis and Prevention, 35, 273-285.
12. Delen, D., Sharada, R. and Bessonov, M. (2006) Identifying Significant Predictors of Injury Severity in Traffic Accidents Using a Series of Artificial Neural Networks. Accident Analysis and Prevention, 38, 434-444.
13. Gelman, A. and Hill, J. (2007) Data Analysis Using Regression and

Multilevel Hierarchical Models. Cambridge University Press, London.

14. Kim, D.G., Lee, Y., Washington, S. and Choi, K. (2007) Modeling Crash Outcome Probabilities at Rural Intersections: Application of Hierarchical Binomial Logistic Models. Accident Analysis and Prevention, 39, 125-134.
15. Abdulhafedh, A. (2016) Crash Frequency Analysis. Journal of Transportation Technologies, 6, 169-180.
16. Blincoe, J., Miller, R., Zaloshnja, E. and Lawrence, A. (2015) The Economic and Societal Impact of Motor Vehicle Crashes, 2010. National Highway Traffic, Washington DC.
17. Park, S. and Lord, D. (2007) Multivariate Poisson-Lognormal Models for Jointly Modeling Crash Frequency by Severity. Transportation Research Record, 2019, 1-6. https://doi.org/10.3141/2019-01
18. Ma, J., Kockelman, K.M. and Damien, P. (2008) A Multivariate Poisson-Lognormal Regression Model for Prediction of Crash Counts by Severity, Using Bayesian Methods. Accident Analysis and Prevention, 40, 964-975.
19. El-Basyouny, K. and Sayed, T. (2009) Collision Prediction Models Using Multivariate Poisson-Lognormal Regression. Accident Analysis and Prevention, 41, 820- 828.
20. Lord, D. and Bonneson, A. (2007) Development of Accident Modification Factors for Rural Frontage Road Segments in Texas. Transportation Research Record, 2023, 20-27.
21. Daniels, S., Brijs, T., Nuyts, E. and Wets, G. (2010) Explaining Variation in Safety Performance of Roundabouts. Accident Analysis and Prevention, 42, 292-402.
22. Malyshkina, N. and Mannering, F. (2010) Markov Switching Multinomial Logit Model: An Application to Accident-Injury Severities. Accident Analysis and Prevention, 41, 829-838.
23. Geedipally, R., Lord, D. and Dhavala, S. (2012) The Negative-Binomial Lindley Generalized Linear Model: Characteristics and Application Using Crash Data. Accident Analysis and Prevention, 45, 258-265.
24. Lord, D. (2006) Modeling Motor Vehicle Crashes Using Poisson-Gamma Models: Examining the Effects of Low Sample Mean Values and Small Sample Size on the Estimation of the Fixed Dispersion Parameter. Accident Analysis and Prevention, 46, 751-766.
25. Oh, J., Washington, S. and Nam, D. (2006) Accident Prediction Model

for Railway-Highway Interfaces. Accident Analysis and Prevention, 38, 346-356.

26. Lord, D. and Miranda-Moreno, F. (2008) Effects of Low Sample Mean Values and Small Sample Size on the Estimation of the Fixed Dispersion Parameter of Poisson-Gamma Models for Modeling Motor Vehicle Crashes: A Bayesian Perspective. Accident Analysis and Prevention, 46, 751-770.
27. Aguero-Valverde, J. and Jovanis, P. (2008) Analysis of Road Crash Frequency with Spatial Models. Transportation Research Record, 2061, 55-63. https://doi.org/10.3141/2061-07
28. Xie, Y. and Zhang, Y. (2008) Crash Frequency Analysis with Generalized Additive Models. Transportation Research Record, 2061, 39-45. https://doi.org/10.3141/2061-05
29. Li, X., Lord, D., Zhang, Y. and Xie, Y. (2009) Predicting Motor Vehicle Crashes Using Support Vector Machine Models. Accident Analysis and Prevention, 40, 1611- 1618.
30. Milton, J.C., Shankar, V. and Mannering, F. (2008) Highway Accident Severities and the Mixed Logit Model: An Exploratory Empirical Analysis. Accident Analysis and Prevention, 40, 260-266.
31. Anastasopoulos, P.C. and Mannering, F. (2009) A Note on Modeling Vehicle Accident Frequencies with Random-Parameters Count Models. Accident Analysis and Prevention, 41, 153-159.
32. Washington, P., Karlaftis, G. and Mannering, F. (2010) Statistical and Econometric Methods for Transportation Data Analysis. 2nd Edition, Chapman Hall, Control and Reporting Center, Boca Raton.
33. Meng, H., Zheng, L. and Qing, M. (2009) Traffic Accidents Prediction and Prominent Influencing Factors Analysis Based on Fuzzy Logic. Accident Analysis and Prevention, 9, 87-92.
34. Abdelwahab, H.T. and Abdel-Aty, M.A. (2002) Artificial Neural Networks and Logit Models for Traffic Safety Analysis of Toll Plazas. Accident Analysis and Prevention, 1784, 115-125. https://doi.org/10.3141/1784-15
35. Riviere, C., Lauret, P., Ramsamy, M. and Page, Y. (2006) A Bayesian Neural Network Approach to Estimating the Energy Equivalent Speed. Accident Analysis and Prevention, 38, 248-259.
36. Chang, L.Y. (2005) Analysis of Freeway Accident Frequencies: Negative Binomial Regression versus Artificial Neural Network.

Accident Analysis and Prevention, 43, 541-557.

37. Cameron, A.C., and Trivedi, P.K. (1998) Regression Analysis of Count Data. Cambridge University Press, Cambridge, UK. https://doi.org/10.1017/CBO9780511814365
38. Abdel-Aty, M. (2003) Analysis of Driver Injury Severity Levels at Multiple Locations Using Ordered Probit Models. Journal of Safety Research, 34, 597-603.
39. Chang, L.Y., and Wang, H. (2006) Analysis of Traffic Injury Severity: An Application of Nonparametric Classification Tree Techniques. Accident Analysis and Prevention, 38, 1019-1027. https://doi.org/10.1016/j.aap.2006.04.009
40. Bham, G., Javvadi, B. and Manepalli, U. (2012) Multinomial Logistic Regression Model for Single-Vehicle and Multivehicle Collisions on Urban U.S. Highways in Arkansas. Journal of Transportation Engineering, 138, 786-797 https://doi.org/10.1061/(ASCE)TE.1943-5436.0000370
41. Glenberg, A. (1996) Learning from Data: An Introduction to Statistical Reasoning. 2nd Edition, Lawrence Erlbaum Associates, Mahwah.
42. Hilbe, J. (2007) Negative Binomial Regression. Cambridge University Press, London. https://doi.org/10.1017/CBO9780511811852
43. Hilbe, J. (2014) Modeling Count Data. Cambridge University Press, London. https://doi.org/10.1017/CBO9781139236065
44. Amoros, E., Martin, J.L., & Laumon, B. (2003) Comparison of Road Crash Incidents and Severity between Some French Counties. Accident Analysis and Prevention, 35, 537-547. https://doi.org/10.1016/S0001-4575(02)00031-3
45. Shankar, N., Milton, J.C. and Mannering, F. (1997) Modeling Accident Frequencies as Zero-Altered Probability Process: An Empirical Enquiry. Accident Analysis and Prevention, 29, 829-837.
46. Lambert, D. (1992) Zero-Inflated Poisson Regression with an Application to Defects in Manufacturing. Technometrics, 34, 1-14. https://doi.org/10.2307/1269547
47. Zhang, J.X., Chang, K.T. and Wu, J.Q. (2008) Effects of Dynamic Effect Model Resolution and Source on Soil Erosion Modeling: A Case Study Using the Water Erosion Prediction Project Model. International Journal of Geographical Information Science, 22, 925-942. https://doi.org/10.1080/13658810701776817

48. Greene, W. (2008) Econometric Analysis. 6th Edition, Prentice-Hall. Upper Saddle River.
49. Wang, H., Zheng, L. and Meng, X.H. (2011) Traffic Accidents Prediction Model Based on Fuzzy Logic. Communications in Computer and Information Science, 201, 101-108. https://doi.org/10.1007/978-3-642-22418-8_14
50. Greene, W. (2012) Econometric Analysis. 7th Edition, Prentice Hall, Upper Saddle River.
51. Savolainen, P., Mannering, F., Lord, D. and Quddus, M. (2011) The Statistical Analysis of Highway Crash-Injury Severities: A Review and Assessment of Methodological Alternatives. Accident Analysis and Prevention, 43, 1666-1676.
52. Mannering, F. and Grosdsky, L. (1995) Statistical Analysis of Motorcyclist: Perceived Accident Risk. Accident Analysis and Prevention, 27, 21-31.
53. Judge, G., Griffiths, W.E., Hill, R.C., Lutkepohl, H. and Lee, T.C. (1985) The Theory and Practice of Econometrics. 2nd Edition, Wiley, New York.
54. Baltagi, B.H. (2011) Econometrics. 5th Edition, Springer, Berlin. https://doi.org/10.1007/978-3-642-20059-5

CHAPTER

9

High-Risk Rural Road Safety Study and Determining the Crash-Reduction Factors for High-Risk Rural-Road Usage

Ranjit Prasad Godavarthy[1], Eugene R. Russell[2]

[1]Upper Great Plains Transportation Institute, North Dakota State University, Fargo, North Dakota, USA

[2]Department of Civil Engineering, Kansas State University, Manhattan, Kansas, USA

ABSTRACT

More than 32,000 motorists are killed on U.S. roads and streets annually, and approximately 54% of the accidents occur on rural roads. In an attempt to address and reduce these fatalities, the current transportation act, the Safe, Accountable, Flexible, Efficient Transportation Equality Act: A Legacy for

Citation: Godavarthy, R. and Russell, E. (2016), "High-Risk Rural Road Safety Study and Determining the Crash-Reduction Factors for High-Risk Rural-Road Usage". *Journal of Transportation Technologies*, **6**, 1-8. doi: 10.4236/jtts.2016.61001.

Users (SAFETY-LU), elevated the Highway Safety Improvement Program (HSIP) to a core program and included a $90,000,000 High-Risk Rural Road Program (HRRRP) to address and significantly reduce traffic fatalities and incapacitating injuries on rural major or minor collectors, and/or rural local roads. While there were many challenges to properly implement the HRRRP in counties, this study provided important information that was needed to identify the predominant crash types on HRRRP-eligible roads and compiled a list of countermeasures for the predominant crash types that were identified on Kansas' high-risk rural roads. For the gathered countermeasures, crash reduction factors (CRFs) were also provided from the literature review, and their values were validated by conducting interviews with Kansas county engineers/officials. This study provided valuable information for the county engineers and local government officials while they worked on improving the safety of high-risk rural roads using HRRRP funds.

Keywords: HRRR, HSIP, Rural Road Safety, Crash Reduction Factors

INTRODUCTION

Rural roads are susceptible to a high rate of crashes, and the crash severities tend to be worse, leading to many fatalities and incapacitating-injury crashes [1] - [3] . The main reason for most of the rural-road crashes is due to the high speed that vehicles travel on rural roads [1] . According to a NHTSA report, among the total fatalities in the United States in 2012, 54% of them were recorded in rural areas even though only 19% of the U.S. population lived in rural areas [4] .

To improve the safety of high-risk rural roads in the United States, the Federal Highway Administration (FHWA) has established the High-Risk Rural Road Program (HRRRP) under SAFETY-LU since 2005 as a mandatory set-aside in the highway-safety improvement program (HSIP) in order to provide funds to improve road deficiencies and to decrease the number of rural-road crashes [5] - [7] . The functional classes of roadways that are eligible for HRRRP funds are rural major or minor collectors, or rural local roads [6] . Further, the eligible roadways must have crash rates that exceed the statewide average for the respective roadway's functional classifications. Only fatal and incapacitating-injury crashes are considered, and their crash rates are determined by using exposure data, such as vehicle miles traveled (VMT), average daily traffic (ADT), lane miles, etc. However, other data-driven methods can be considered for determining the eligible roadways for HRRRP funds [6] . Projects that are eligible for HRRRP include construction and operational projects, and not projects dealing with data-system improvement, improving enforcement, and increasing educational efforts [8] .

Using HRRRP to improve safety for high-risk rural roads can be challenging. Some common difficulties that state experience while improving the safety of rural roads by using HRRRP include gathering crash data and exposure data to identify locations for potential improvements; selection criteria for the best projects; and coordination among federal, state, and local transportation partners [8] .

Gathering crash data can be a huge challenge because some counties do not maintain crash data. If the crash data are available, determining the accurate crash location and the availability of exposure data is questionable [8] . Therefore, providing useful information, such as identifying the significant crash types on the HRRRP-eligible roads, having guidelines to select an efficient project for maximum safety improvements, and providing the needed information to improve the usage of HRRRP funds, to county officials can lead to safe rural roads.

Rural-safety needs generally far outstrip the available HRRRP funds, along with the other funds that are available to many local jurisdictions; therefore, road deficiencies should be prioritized such that the limited funds that are available can provide the maximum, or most cost-effective, safety return per dollar spent. While funds and resources are available to improve the safety of rural roads, the county engineers and local jurisdictions need more information regarding the predominant crash types that are occurring on high-risk rural roads, all possible countermeasures for each crash type, and crash-reduction factors which are specific and appropriate for rural roads.

Therefore, the study's objectives are as follows:

1). Describe a procedure to select an efficient project/countermeasure for an identified high-risk rural-road location;
2). Identify the predominant crash types for high-risk rural roads;
3). Compile countermeasures for all the predominant crash types;
4). Gather the crash-reduction factors for each countermeasure, and validate them by interviewing county engineers.

These objectives are achieved by studying Kansas' crash database and by contacting Kansas' county engineers to validate the crash-reduction factors.

STUDY METHODOLOGY

The state of Kansas was considered for this study and for demonstrating the study methodologies. The study's initial objective was to identify the predominant crash types for high-risk rural roads. To determine the crash types that were predominant on high-risk rural roads, an analysis of the available crash database should be conducted. Kansas' crash database was studied to determine the predominant crash types for high-risk rural roads. Later, the crash reduction factors (CRFs) of all possible countermeasures for the predominant crash types were gathered from the literature. These CRFs were validated and modified by conducting interviews with county engineers in Kansas.

The functional classes of roads that are eligible for operational improvements through HRRRP funding include rural major or minor collectors, or rural local roads. Further, the fatal or incapacitating-injury crash rate for these roads should be higher than the statewide average per functional classification in order to be eligible for HRRRP funding to improve the eligible rural roads' safety. Therefore, fatal and incapacitating-injury crashes in Kansas were analyzed for the functional road classes, such as rural major collectors, rural minor collectors, and rural local roads, to identify the most significant crash types.

Improving a High-Risk Rural-Road Location Using HRRRP

To determine that a rural road was eligible for HRRRP funds, the crash (fatal or incapacitating) rate for a segment of an eligible rural road should be more than the statewide average. When a section of rural road had a crash rate higher than the statewide average, all the crashes that occurred at that location were studied to determine the predominant crash types for the road segment. If a certain crash type was predominant, then the segment's safety could be improved by adopting suitable countermeasures. Various countermeasures were available for each crash type, and every countermeasure was assigned a CRF which was derived from research and case studies that determined how much of a crash reduction can be expected after installing a countermeasure at a location with prior accidents. CRFs are expressed as a percentage or decimal which adds up to 100% or 1.00. The CRF gives the number of crashes that can be reduced by the countermeasure by multiplying the factor by the "before" number of crashes. As shown in Figure 1, an effective countermeasure, among the many available countermeasures for a crash type, can be determined by knowing each countermeasure's CRF value and then conducting a cost-benefit analysis.

KANSAS CRASH DATA ANALYSIS

The Kansas Crash Analysis Reporting System (KCARS) database was used to analyze the crashes on rural roads that were eligible for HRRRP in the state of Kansas [9] . The KCARS database included crash records reported by law-enforcement agencies following the state's reporting threshold, i.e., an accident involving a fatality, an injury, or property damage of at least $1000 in aggregate. The KCARS database was analyzed for the years of 2002 to 2006 to identify the crash types for the following functional classes of roads: rural major collectors, rural minor collectors, and rural local roads. Only crash-severity classifications of fatal and incapacitating injuries were considered for the analysis. Together, these crashes were considered serious crashes.

From the 5-year (2002-2006) crash-data analysis in Kansas, a total of 2533 crashes with fatalities or incapacitating injuries were identified for the functional roadway classes of rural major collectors, rural minor collectors, and rural local roads. Among these crashes, most of them occurred on rural major collectors (1364 crashes) and rural local roads (1023 crashes), and very few occurred on rural minor collectors (146 crashes). These crashes were analyzed and categorized based on accident severity, accident class, manner of collision, fixed-object type, and accident location. Categorizing the crash data can help understand the predominant crash types on the rural roads that are eligible for HRRRP.

Figure 2 summarizes the crash categorization based on accident class for the three functional road classes that were studied. It can be understood, from Figure 2, that overturned crash, a crash involving a collision with another motor vehicle, and a crash involving a collision with a fixed object are major contributors. It has also been observed that 21% of the total overturned crashes and 20% of the total crashes involving a collision with a fixed object are further categorized as run-off-road crashes in the KCARS database. Crashes involving a collision with another motor vehicle are further categorized and summarized in Figure 3. It can be observed from Figure 3 that angle-side impact crashes are a major share of the collision-with-another-motor-vehicle crashes. The other two major types of crashes involving a collision with another motor vehicle are head-on crashes and rear-end crashes. While most angle-side impact crashes occur at rural intersections, this crash type is not in the study's scope.

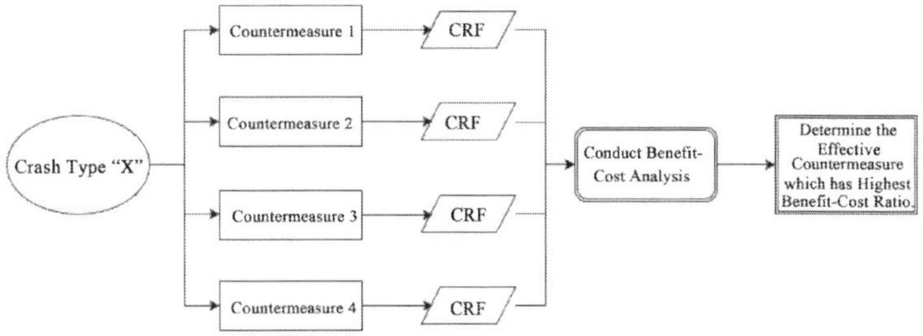

Figure 1. Procedure for determining an effective countermeasure for maximum safety benefits.

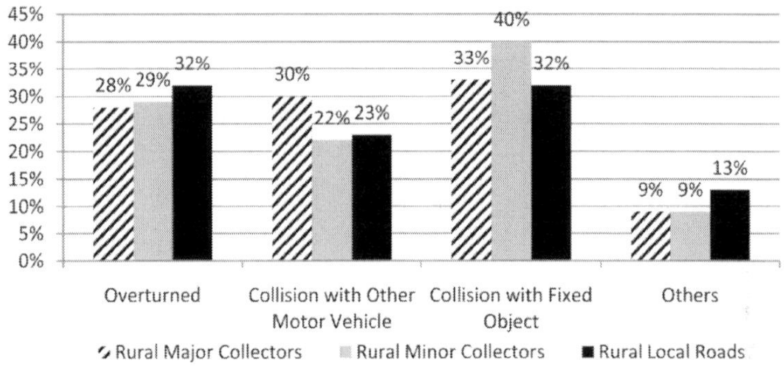

Figure 2. Crash categorization based on accident class.

This study primarily focuses on improving the safety for a segment of rural roads excluding the intersections because rural-intersection safety can be an entirely different kind of study and has been well researched. More information about intersection safety, various possible countermeasures available to enhance the intersection safety, and crash-reduction factors can be found in a FHWA document, "Safety at Unsignalized Intersections", and a Kansas Department of Transportation report, "Mitigating Crashes at High-Risk Rural Intersections with Two-Way Stop Control" [10] [11] . Crashes involving a collision with a fixed object are further categorized and presented in Figure 4. It can be understood, from Figure 4, that fixed-object crash types, such as ditch and tree, are comparatively higher among many fixed-object crashes. While there are many kinds of fixed-object crashes identified, the common reason for these crashes is running off the road.

Therefore, from the five-year crash analysis of HRRRP-eligible roads in Kansas, the predominant crash types are fixed-object crashes, run-off-road crashes, overturned crashes, rear-end crashes, and head-on crashes.

The Federal Highway Administration (FHWA) has compiled various countermeasures and the crash-reduction factors (CRFs) for each crash type from all known, reliable sources into a desktop reference [12] . The CRF determines how much of a crash reduction can be expected after installing some countermeasure at a location with prior accidents. All possible countermeasures and their CRFs for the five predominant crash types that are identified above are gathered from the FHWA report "Desktop Reference for Crash Reduction Factors" and listed in Table 1 and Table 2 [12] .

The CRFs gathered from the FHWA report [12] are average estimates based on numerous studies; however, they can be of value when estimating the cost effectiveness and/or ranking, as well as prioritizing, for a group of countermeasures. Further, the CRFs gathered from the FHWA report [12] are based mostly on the expected accident reductions for state highways, hence these CRFs may or may not be highly reliable for low-volume county roads. It would be best for a given county to keep track of all accidents before and after various safety countermeasures are implemented and to develop its own CRFs.

This study attempted to determine the CRFs for Kansas' high-risk rural roads by interviewing the county engineers to obtain the county's unique CRF values and also to inquire about if the CRFs are closer to the CRFs compiled in the FHWA report [12] . A total of 50 counties in the state of Kansas were contacted and given survey forms to gather the CRF values from Kansas' county engineers for all the countermeasures regarding the five predominant crash types that were identified for HRRRP-eligible roads. Among the 50 counties contacted, 11 counties responded, making the response rate 22%. A few more county officials responded by e-mail saying that they did not have any accurate CRFs. Results from the returned survey forms are summarized in Table 1 and Table 2. About 50% of the countermeasures had CRFs that matched the FHWA's reference report, and the other half of the countermeasures had different CRF values. Also, when the CRFs from the survey results matched the CRFs from the FHWA report, it was mainly due to the fact the FHWA repot had a wide range of CRF values.

Table 1. Countermeasures and CRFs for fixed-object, run-off-road, and overturned crashes.

Countermeasure Average CRF from county engineer's survey (%)	**CRFs**			
	CRF from FHWA report (%) [12]	**% of survey responses matching the FHWA CRFs**	**CRF from the surveys in the FHWA's CRF range? (Yes/No)**	
Fixed-object crash type				
Widen shoulder (paved)	34.8	15 - 62	90%	Yes
Widen shoulder (unpaved)	23.1	13 - 43	80%	Yes
Remove or relocate fixed objects outside the clear zone	67.6	40 - 100	60%	Yes
Widen the clear zone	28.4	13 - 44	50%	Yes
Improve pavement friction (grooving)	22	19 - 36	77%	Yes
Install edge-line markings	59.6	59 - 66	100%	Yes
Improve guardrail	16.17	18 - 21	67%	No
Install guardrail (as shield for rocks and posts)	81.1	100	80%	No
Install impact attenuators	46.1	69	50%	No
Flatten horizontal curve	61.5	68 - 87	77%	No
Flatten side slopes	50.7	62	80%	No
Improve pavement friction (overlay)	35	43	80%	No
Improve pavement friction (resurface with open-graded mix)	75	93	80%	No
Install or upgrade curbing	39.6	50	67%	No
Run-off-road crash type				
Install guardrail (at embankment)	22.8	7 - 47	67%	Yes
Flatten horizontal curve	75.2	79 - 90	89%	Yes
Flatten side slopes	17	10 - 24	70%	Yes
Widen lane	25.6	12 - 49	80%	Yes
Widen shoulder (paved)	35.4	60 - 16	73%	Yes

Widen shoulder (unpaved)	23.7	13 - 43	70%	Yes
Improve pavement friction (groove shoulder)	27.8	27 - 41	80%	Yes
Install shoulder rumble strips	23.1	7 - 41	77%	Yes
Install edge-line markings	25.3	25 - 30	100%	Yes
Improve guardrail	26.7	28 - 41	67%	No
Improve super-elevation	44.6	50	89%	No
Remove or relocate fixed objects outside the clear zone	64.2	71	77%	No
Install curve advance-warning signs	28.1	30	80%	No
Install delineators (general)	33.7	34	80%	No
Improve pavement friction (overlay)	23	28	80%	No
Overturned crash type				
Improve guardrail	31.2	27 - 41	80%	Yes
Install guardrail (at bridge)	32.8	32 - 41	90%	Yes
Flatten horizontal curve	43.6	24 - 73	90%	Yes
Remove or relocate fixed objects outside the clear zone	35.1	42 - 44	77%	No
Improve pavement friction (grooving)	35.2	35 - 54	77%	No

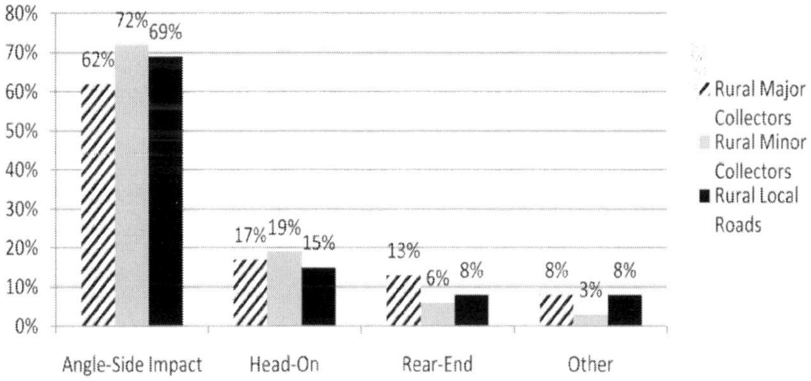

Figure 3. Crash categorization based on the manner of colliding with another motor vehicle.

Table 2. Countermeasures and CRFs for rear-end and head-on crashes.

Countermeasure	**CRFs**			
	Average CRF from the county engineers' survey (%)	**CRF from the FHWA report (%) [12]**	**% of survey re-sponses match-ing the FHWA CRFs**	**CRF from the surveys in the FHWA's CRF range? (Yes/No)**
Rear-end crash type				
Flatten horizontal curve	37	24 - 73	75%	Yes
Increase the number of lanes	32.5	32 - 53	75%	Yes
Improve pavement friction (grooving)	35	35 - 54	63%	Yes
Install acceleration/decelera-tion lanes	64.5	75	77%	No
Install channelized lane	70.4	93	75%	No
Install lane-assignment signs	8.1	10	77%	No
Improve pavement friction (overlay)	11.4	12	70%	No
Install edge-line markings	35.3	45 - 50	67%	No
Implement maintenance and bituminous overlay	18.5	21	60%	No
Head-on crash type				
Widen lane	36.2	5 - 70	70%	Yes
Widen shoulder (paved)	41.7	15 - 75	70%	Yes
Improve pavement friction (overlay)	31.5	19 - 61	70%	Yes
Install centerline rumble strips	41.3	21 - 68	77%	Yes
Widen bridge	37	40 - 50	80%	No
Flatten horizontal curve	47.8	64 - 67	70%	No
Increase the number of lanes	36	38 - 53	77%	No
Install shoulder bus lanes	38.9	50	77%	No
Install curve advance-warn-ing signs	24.2	29	80%	No
Install delineators (general)	58.6	67	80%	No

Improve pavement friction (curve overlay)	68	86	70%	No
Improve pavement friction (resurface with deicing additives)	21.8	31	67%	No
Improve pavement friction (resurface with open-graded mix)	71.1	90	77%	No
Pave shoulder	61.6	86	56%	No
Install no-passing line	37.7	40	77%	No

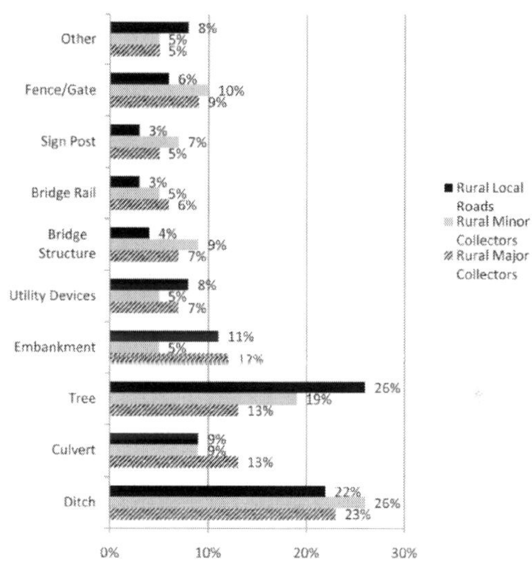

Figure 4. Crash categorization based on a fixed-object type.

Further, having a range of CRF values for a countermeasure might lead to inaccurate safety-benefit prediction for high-risk rural roads. Therefore, it would be very efficient to have a precise CRF value for efficient safety benefit prediction of countermeasures on high-risk rural roads. This study presents precise CRF values for various countermeasures that can be readily applicable for HRRRP-eligible roads because these CRF values are derived by the crash reduction observed by implementing the countermeasures on high-risk rural roads. Therefore, the CRFs gathered by surveying the Kansas county engineers can be more reliable and accurate when predicting the safety effectiveness of the compiled countermeasures with high-risk rural roads.

CONCLUSIONS AND RECOMMENDATIONS

This study analyzed the crashes on high-risk rural roads and focused on improving these roads' safety by providing more information and tools for the FHWA High-Risk Rural Road Program (HRRRP). The state of Kansas was selected for analyzing and demonstrating the study methodologies. Fatal and incapacitating-injury crashes on Kansas' rural major collectors, rural minor collectors, and rural local roads were analyzed for five- year crash data, and the predominant crashes types that were identified were fixed-object crash, run-off-road crash, overturned crash, rear-end crash, and head-on crash. Various possible countermeasures and the CRFs for each predominant crash type were gathered from the literature, and the CRF values were validated for high-risk rural-road usage by conducting interviews with Kansas county engineers/officials. The predominant crash types might be different for high-risk rural roads in other U.S. states, but basic crash-data analysis for the rural roads would be sufficient to identify the predominant crash types.

From the 50 Kansas counties contacted, only 12 counties responded, making the response rate 22%. The CRF values obtained by Kansas county interviews were different from the CRF values available through the FHWA's report [12] for most of the countermeasures, proving that the countermeasures' crash-reduction factors could be different for rural major collectors, rural minor collectors, and rural local roads. Some counties officials who did not respond with their CRF values, provided feedback that, they did not have any accurate CRFs because they did not maintain a record of accidents over time. However, if before-after crash results are recorded by county engineers when various countermeasures are installed, then, over time, reliable CRFs could be developed for any given county or U.S. state. Reliable CRFs can be a valuable tool when setting priorities for needed safety improvements. The authors believe that, although CRFs should not be considered as an absolute, accurate value, they are better than not having any idea what safety benefit will result from a given improvement and have a value in prioritizing several possible improvements.

Updated countermeasure CRFs for this study's identified predominant crash types can be useful for county engineers and local government officials to improve the safety of high-risk rural roads with available funding programs, particularly HRRRP.

ACKNOWLEDGEMENTS

The funding for this study is provided by Kansas State University Transportation Center and Kansas Department of Transportation.

REFERENCES

1. Aarts, L. and Schagen, I.V. (2006) Driving Speed and the Risk of Road Crashes: A Review. Accident Analysis and Prevention, 38, 215-224. http://dx.doi.org/10.1016/j.aap.2005.07.004
2. Dissanayake, S. and Perera, L (2011) A Survey Based Study of Factors Related to Older Driver Highway Safety. Journal of Transportation Safety & Security, 3, 77-94. http://dx.doi.org/10.1080/19439962.2010.537437
3. Perera, L. and Dissanayake, S. (2010) Contributing Factors to Older-Driver Injury Severity in Rural and Urban Areas. Journal of Transportation Research Forum, 49, 5-22.
4. National Highway Traffic Safety Administration (2014) Traffic Safety Facts 2012. U.S. Department of Transportation, National Highway Traffic Safety Administration, Washington DC.
5. Federal Highway Administration (2005) Safe, Accountable, Flexible, Efficient Transportation Equity Act: A Legacy for Users. FHWA, Office of Legislation and Intergovernmental Affairs, Washington DC.
6. Federal Highway Administration. (2006) High Risk Rural Roads Interim Guidance. http://safety.fhwa.dot.gov/safetealu/memos/memo051906.cfm
7. Federal Highway Administration (2008) Highway Safety Improvement Program. Publication FHWA-2008-0009, U.S. Department of Transportation, Washington DC.
8. Chandler, B. and Anderson, R. (2010) Implementing the High Risk Rural Road Program. Publication FHWA-SA-10-012. Federal Highway Administration, Washington DC.
9. Kansas Accident Reporting System (2009) Kansas Department of Transportation, U.S. Department of Transportation.
10. Federal Highway Administration. Safety at Unsignalized Intersections. http://safety.fhwa.dot.gov/intersection/
11. Russell, E.R. and Godavarthy, R.P. (2010) Mitigating Crashes at High-Risk Rural Intersections with Two-Way Stop Control. Report No. K-TRAN: KSU-06-4, Kansas Department of Transportation.
12. Bahar, G., Masliah, M., Wolff, R. and Park, P. (2007) Desktop Reference for Crash Reduction Factors. Report No. FHWA-SA-07-015, U.S. Department of Transportation, Washington DC.

SECTION III
PREDICTION OF THE PUBLIC TRANSPORT

CHAPTER 10

Prediction of Civil Aviation Passenger Transportation Based on ARIMA Model

Xinxin Tang[1], Guangming Deng[1,2]

[1]College of Science, Guilin University of Technology, Guilin, China

[2]Institute of Applied Statistics, Guilin University of Technology, Guilin, China

ABSTRACT

The passenger transportation, as an important index to describe the scale of aviation passenger transport, prediction and research, can let us understand the future trend of the aviation passenger transport, according to it, the airline can make corresponding marketing strategy adjustment. Combining with the knowledge of time series let us understand the characteristics of

Citation: Tang, X. and Deng, G. (2016), "Prediction of Civil Aviation Passenger Transportation Based on ARIMA Model". *Open Journal of Statistics*, **6**, 824-834. doi: 10.4236/ojs.2016.65068.

passenger transportation change, the R software is used to fit the data, so as to establish the ARIMA(1,1,8) model to describe the civil aviation passenger transport developing trend in the future and to make reasonable predictions.

Keywords: Passenger Transportation, ARIMA Model, Seasonal Trend, Forecast

INTRODUCTION

In recent years, with the improvement of people's living, travel distance extension, travel mode is no longer confined to the car and train traffic. And taking into account the cost of time, more and more people choose the aircraft as means of transportation to travel. Some scholars have done research on the characteristics of China's air transport development [1] that with the development of the national economy it has a certain period of regularity. As China's comprehensive national strength, the aviation market will also have the potential to develop more. Passenger demand forecasting value of civil aviation is the key factor to determine the size of the passenger transport market. This article will focus on the future development of civil aviation passenger transport prospects.

Now prediction methods that have complete theoretical basis mainly include: regression analysis, time series analysis, input-output method, inductive method and Markov chain prediction method. At present, China has already done the research and forecast of aviation passenger traffic. In terms of influence factors, Wang Cui (2008) [2] from the perspective of macroeconomic and microeconomic analysis, found that there are many factors impact the air passenger volume and the factors include non quantifiable factors. Deng Jiejun and Luo Li (2006) [3] , based on the relationship between social economic development and civil aviation passenger demand, use the method of multiple linear regression and GMDH to predict and analyze the demand of civil aviation passenger transport. In methodology, Chen Lihua (2003) [4] , in accordance with the method of regression analysis, taking domestic and international airline passenger turnover as a result of variables, predicts the total turnover of passengers. Wang Xiaoguang et al. (2008) [5] proposed a model which based on artificial neural network and the combination of the polynomial model to forecast the volume of passenger. Pan Ling et al. (2014) [6] established the product season model to forecast the civil aviation passenger traffic volume from December 2013 to December 2014. Kong Jianguo et al. (2014) [7] used the wavelet analysis theory to forecast the demand of civil aviation passenger transport.

Because there are many factors influencing the air passenger volume, we need in-depth study of the internal and external environment of the aviation industry. The influence factors include the non quantifiable factors which are not easy to treatment, so this paper selects the self factor model focuses on its internal rules. In economic forecasting process, the ARIMA model (i.e., autoregressive moving average model) not only considers the economic phenomena in time series dependence, but also considers the interference of random fluctuations. On this problem of short-term trend forecast for the economic operation, this model has the high accuracy rate. The ARIMA model is widely used in economic research. Because of many factors such as holidays and climate factors, the aviation passenger transport industry has the market volatility. The passenger volume can be regarded as a stochastic time series formed with the passage of time. By analysing the time series is stochastic, stationary and seasonal factors or not, we can use the ARIMA model to fit civil aviation passenger transport. Therefore, this paper will use the time series analysis of the ARIMA model and R software to fit the data to achieve passenger transport turnover forecast. Air passenger traffic turnover is a composite index that can describe air transport enterprises in a certain period of time. It is a composite index of transport volume and distance. Air passenger traffic turnover is not only an important index of civil aviation enterprises, but also one of the main indicators of the national assessment of air transport enterprises. Therefore, this paper will study the problem of air passenger transport turnover forecast.

THEORETICAL BASIS

ARIMA Model

In the domestic and foreign well-known materials or works, modeling method of time series is basically uniform that the steady but autocorrelation sequence to establish ARMA model and the non-stationary time series (with unit root) to establish ARIMA(p,d,q) model [8] .

The ARIMA(p,d,q) model includes the autoregressive model AR(p), moving average model MA(q) and the mixed autoregressive moving average model ARMA(p,q) of these three cases [9] .

The first one, the AR(p) model corresponding to the algebraic expression is:

$$y_t = c + \alpha_1 y_{t-1} + \alpha_2 y_{t-2} + \cdots + \alpha_p y_{t-p} + \varepsilon_t \tag{1}$$

The second, the MA(q) model of the corresponding algebraic expression is:

$$y_t = c + \varepsilon_t + \theta_1\varepsilon_{t-1} + \theta_2\varepsilon_{t-2} + \cdots + \theta_q\varepsilon_{t-q} \qquad (2)$$

Third, the ARMA(p,q) model corresponds to the logarithmic expression is:

$$y_t = c + \alpha_1 y_{t-1} + \alpha_2 y_{t-2} + \cdots + \alpha_p y_{t-p} + \varepsilon_t + \theta_1\varepsilon_{t-1} + \theta_2\varepsilon_{t-2} + \cdots + \theta_q\varepsilon_{t-q} \qquad (3)$$

If y_t is not smooth, but $W_t = \Delta^d y_t = (1-L)^d y_t$ is smooth, then integrating the ARMA model to be ARIMA(p,d,q) for the use of a smooth process W_t to replace the position of the unstable y_t in the ARMA model.

Among them, both α and θ are the parameters to be estimated and L is the lag operator.

STL Decomposition Method

The STL [10] is a time series decomposition method with robust local weighted regression as the smoothing method. And the Loess method is a local polynomial regression fitting, which is a common method for smoothing the two dimensional scatter points. It combines the simplicity of the traditional linear regression and the flexibility of the nonlinear regression. When you want to estimate a response variable value, it is first to get a subset of data from the forecast variables nearby, then use linear regression or quadratic regression on the subset regression. The weighted least squares method is adopted, which is close to the estimated value of the point of greater weight. Finally, using the local regression model to estimate the value of the response variable. The whole fitting curve is obtained by the method of point by point operation. This method has a lot of advantages. For example, it can handle any type of seasonal data, the quarter component can be allowed to change over time, it can control the rate of change, and has better robustness to outliers. The disadvantage of this method is that it is only applicable to the additive model.

The STL decomposition method corresponds to the logarithmic expression is:

$$\begin{cases} S_v = S_v^{(k+1)} \\ T_v = T_v^{(k+1)} \\ R_v = Y_v - S_v - T_v \end{cases} \qquad (4)$$

Among them, Y_v as the raw data, S_v is the trend of low pass filtering of time series data, Y_v indicates trend smoothing and R_v is the final result obtained by the STL decomposition method.

SOURCE AND PRETREATMENT OF DATA

This article studies on the index (civil aviation passenger transport volume [11]) data from the main indicators of transportation and production, which be downloaded from the Civil Aviation Administration of China. This paper studies the data for the monthly data from January 2010 to August 2015. Through the R software to draw the timing diagram, as shown in Figure 1.

It can be seen from Figure 1 that the passenger turnover has a strong seasonal cycle, in August each year to reach a significant peak and February is trough. And it can be seen that the turnover of passenger traffic has been increasing in accordance with a relatively stable trend year by year.

The change of seasons can make the time series of passenger turnover change regularly, which is the seasonal periodicity seen in Figure 1. There are many reasons for the seasonal changes, such as climate, holidays and so on. Through seasonal adjustment which can eliminate seasonal effects in the sequence, it to be able to more clearly reveal the trend of the time series [12] . In order to reflect the essential attribute of the objective economy more accurately, we need to take some methods to eliminate and adjust the seasonal variation factors in time series.

From January 2010 to August 2015 passenger turnover data into R software, based on the Loess method to do seasonal trend decomposition. This trend decomposition method is STL decomposition method, which is a time series decomposition method with generality and robustness. Thus it gets the seasonal trend decomposition map (it is shown in Figure 2).

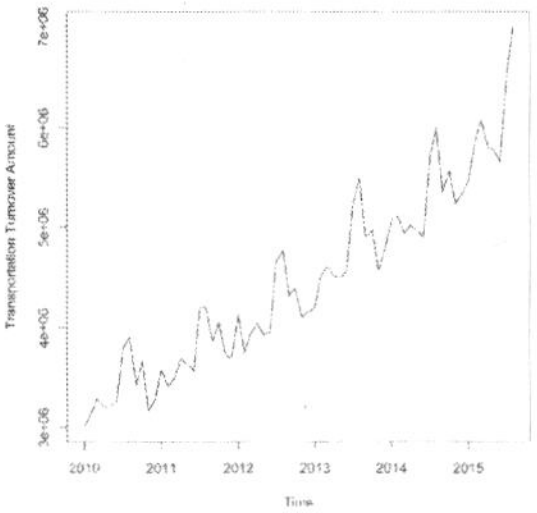

Figure 1. Chinese civil aviation passenger transportation time sequence diagram.

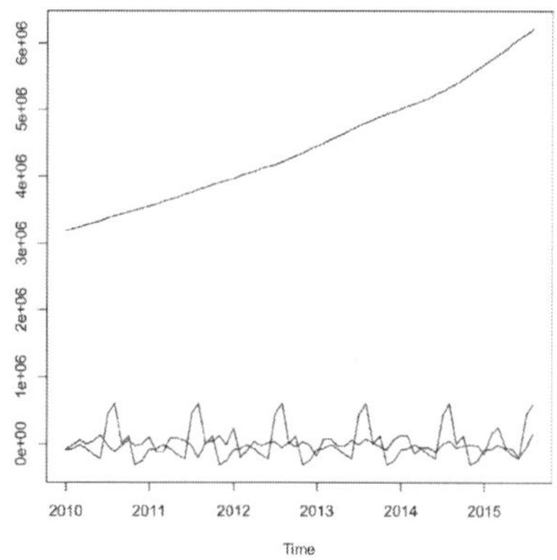

Figure 2. Seasonal trend decomposition.

Table 1. Seasonal variation of each month.

Month	Seasonal variation value
January	−90,584.3
February	−70,172.9
March	−17,375.3
April	−71,211.5
May	−156,815
June	−216,210
July	445,438.6
August	606,397.1
September	7611.507
October	116,642.3
November	−306,071
December	−247,649

As can be seen from Figure 2, the data is divided into three parts, the trend, season and redundancy, in which the various parts of the seasonal variation of the specific data as shown in Table 1. The upward curve of the top of Figure 2 is the development trend of passenger turnover, and the information of the time sequence diagram is consistent with it. Table 1 shows that each year in January is the trough of passenger turnover, in August is the peak of the turnover of passenger traffic.

After understanding the basic trend of the civil aviation passenger transport, in order to effectively predict the future trend of development, we need to fit the passenger transport turnover curve. First, we need to remove the seasonal changes in the time series. And then we make the remaining part of the data for ADF test.

When the ADF level is 0.01, 0.05 and 0.1, the critical value of them is −3.99, −3.45 and −3.13 respectively (the smaller the value of the more significant). And the test results show that the test statistic value of the original sequence is 0.1015, there is no sufficient evidence to reject the original hypothesis. It can not say that the original sequence is smooth. Then after the first order difference of the original data, the test statistic obtained is −5.6418, which is significant at 0.01 level, that is to reject the original hypothesis. Therefore, before the construction of the ARMA model to the data for the first order difference.

CONSTRUCTION OF MODEL

Through the R software output the timing diagram of data (it is shown in Figure 3), although it is the removal of seasonal change but still retain most of the sequence information of the original data. The next step is building ARIMA model.

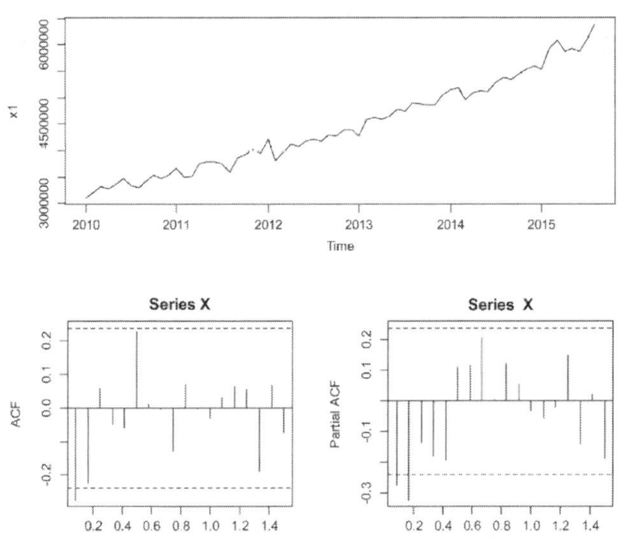

Figure 3. Remove timing map data seasonally (on), after the first order difference of ACF (lower left) and PACF (lower right).

For the ARIMA(p,d,q) model identification, it is required to determine the three parameters of p, d, q. The identification of parameter d has been completed, through the first order difference, the sequence is smooth. So we think the time series is a single integer sequence and it can construct ARMA model after the first order difference. Under normal circumstances, it can be used for the self correlation function ACF map and the partial autocorrelation map PACF map to initially identify p and q. But it is very difficult to see the order in the ACF diagram and PACF diagram. In this paper, we introduce the concept of Bayesian information criterion. BIC is developed from the Bayesian perspective, which is similar to the maximum probability of a posteriori model, and a priori model has a uniform distribution in all models [13] .

The expression formula of BIC is as follows:

$$\mathrm{BIC} = -2\ln\left(L\right) + \ln\left(n\right) \times k \tag{5}$$

Among them, L is the maximum likelihood, n represents the number of data, K said the number of variables in the model.

The BIC criterion is used to describe the loss of information after using a certain model relative to the actual situation. So the smaller the BIC value indicates that the model fitting effect is better. According to the BIC criterion to judge that for different q and p display the corresponding BIC value, the results shown in Figure 4.

As can be seen from Figure 4, the model that minimizes the value of BIC is ARMA(1,8).

It uses ARMA(1,8) model to fit the data which removed seasonal variation (already through the first order difference) and points out the value of p [14] when the lag for the Ljung-Box test is 1 - 30, ACF diagram and PACF diagram is shown in Figure 5.

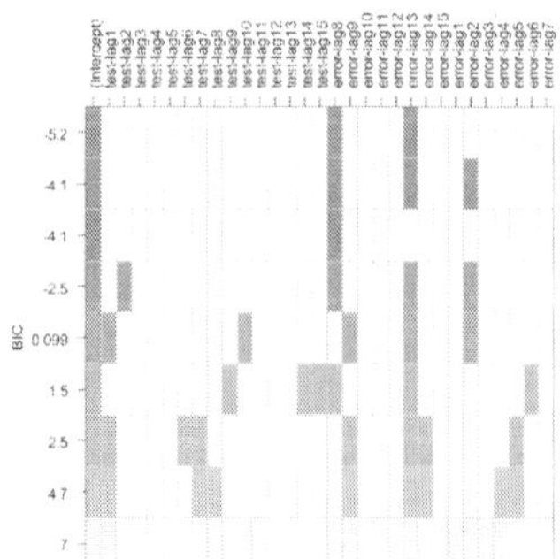

Figure 4. For different p and q shows the corresponding BIC value.

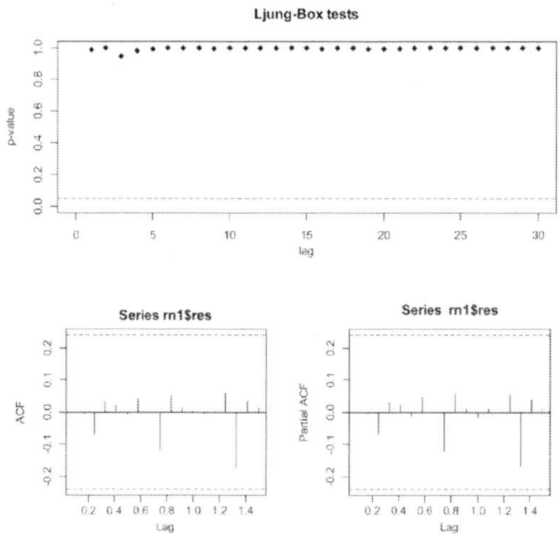

Figure 5. ARMA fitting with the data residuals of the Ljung-Box test of the p value (on), ACF (lower left) and PACF (lower right).

The null hypothesis of Ljung-Box test is that the sequence is independent, for the observed value of p which is not related to should be very large. If the value of p is small that it may be correlated. From Figure 5, we can see that the residual are greater than 0.8, so the residual is white noise and the fitting effect is good. Therefore the model of fitting civil aviation passenger transport turnover is ARIMA(1,1,8), the formula is as follows:

$$x_t = 43691.309 + 0.1182x_{t-1} + \varepsilon_t - 0.7299\varepsilon_{t-1} - 0.5475\varepsilon_{t-2} + 0.3216\varepsilon_{t-3} - 0.0388\varepsilon_{t-4} + 0.0874\varepsilon_{t-5} + 0.3866\varepsilon_{t-6} - 0.3806\varepsilon_{t-7} + 0.0104\varepsilon_{t-8} \quad (6)$$

In the upper representation, x_t is the sequence of the first order difference.

FORECAST RESULT AND ANALYSIS

According to the fitting results of the sequence, the ARIMA(1,1,8) model of civil aviation passenger turnover has passed the test, and the fitting effect is ideal. It can be used to forecast the change of passenger turnover in the future. Using R software to predict the value of passenger turnover in the next 6 months, it draws a forecast figure as shown in Figure 6.

Figure 6 shows the range of passenger traffic volume prediction within the range of error allowed. On the basis of the predicted value of the air passenger turnover, it adds up to the seasonal variation of each month (see Table 1), that is to get the civil aviation passenger transport turnover from the next six months, as shown in Table 2.

As can be seen from the Table 2, the next six months the amount of air passenger traffic to maintain a rising trend, which is affected by seasonal fluctuations will still fluctuate, but the prospects for the airline passenger market is still very impressive. It is worth mentioning that, with the original data of seasonal variation by the seasonal trend decomposition is different, the volume of passenger transport will occur a slightly lower expected value in November 2015, the passenger turnover may be declined in this period of time, but this does not affect the rising trend of the total passenger traffic volume. After comparing with the real data, the prediction error is maintained at about 1%, and the predicted results are in line with the expected results. Same as the predicted results, there is a lower value in November and the overall trend is still maintained steady growth.

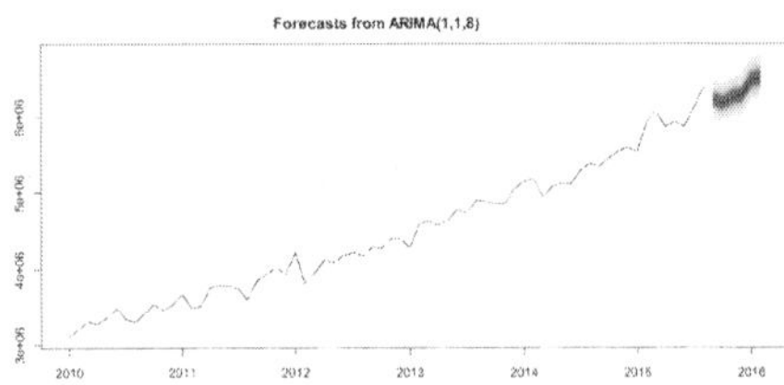

Figure 6. ARIMA(1,1,8) prediction map.

Table 2. Comparison between the predicted and true value of passenger transportation.

Time	Predicted Value	Real Value
2015-9	6,233,218	6,152,663
2015-10	6,296,824	6,420,679
2015-11	5,948,463	5,892,072
2015-12	6,034,594	6,077,000
2016-1	6,387,065	-
2016-2	6,448,524	-
2016-3	6,513,871	-
2016-4	6,530,472	-
2016-5	6,515,186	-
2016-6	6,525,991	-
2016-7	7,257,721	-
2016-8	7,488,643	-

This paper draws a line chart of the passenger turnover amount, as shown in Figure 7. From Figure 7, we can clearly see that the predicted value and the true value is very close, the model has a high prediction accuracy.

CONCLUSIONS AND RECOMMENDATIONS

Time series analysis is performed by the civil aviation monthly passenger volume data from January 2010 to August 2015, this article established the ARIMA(1,1,8) model. On the basis of the ARIMA model to predict the trend of passenger turnover from next 6 months, the fitting effect of this model is better. The following conclusions are obtained by the ARIMA model of passenger turnover:

First, as China's comprehensive national strength and people's living standards to be better, air passenger volume will maintain a steady upward trend, it will officially break the mark of 6 ten thousand person-kilometers. And air passenger market will become bigger.

Second, the passenger turnover has a certain seasonal fluctuation, the main reasons of which include holidays, climate and so on. For example, summer (that is August) is the peak of the air passenger.

Figure 7. Comparison between the predicted and true value of passenger transportation.

The premise of rational planning is a clear understanding of the needs of the future. Only by accurately predicting the trend of air passenger turnover, airlines can reasonably adjust the manpower and material resources in the passenger transport market. In the face of the change of civil aviation passenger volume and passenger market on the future, the airlines should make strategic adjustment decisions, such as good routes and configuration, flight attendant recruitment work and so on. These adjustments will improve the operational capacity and service quality of airlines.

REFERENCES

1. Fan, J. (2003) Research on the Strategy for the Development of China's Air Transportation. Southwest Jiao Tong University, Chengdu.
2. Wang, C. (2008) Prediction of Civil Aviation Passenger Traffic Volume on Grey Theory and RBF Neural Network. Beijing Jiaotong University, Beijing.
3. Deng, J.J. and Luo, J. (2006) Demand Forecasting and Analyzing of Civil Aviation Passenger Transport Markets based on GMDH. Soft Science, 20, 35-38.
4. Chen, L.H. (2003) Analysis and Forecast of Civil Aviation Transport Market. Journal of Shanghai Jiaotong University, 37, 623-625.
5. Wang, X.G. and Zhou, H. (2008) Combinatorial Forecasting Model Based on BP Neural Network in Air Passenger Capacity. Transactions of ShenYang Ligong University, 27, 82-85.
6. Zhu, H.Q. and Pan, L. (2014) An Empirical Study of Civil Aviation Passenger Traffic Volume Based on the Seasonal Product Model. China Market, 6, 115-126.
7. Kong, J.G. and Xiao L. (2014) Application of Wavelet Analysis Based on MATLAB in the Forecasting of Passenger Transport Capacity of Civil Aviation. China Science and Technology Information, 16, 40-41.
8. Tian, Y.F. and Jiang, Q.Q. (2015) An Important Problem of Time Series Analysis and Its Solution. Statistics and Decision, 6, 87-88.
9. Li, Y.Y., Liu R. and Ding, W.D. (2013) Statistical Analysis and Application of EViews. Electronic Industry Press, Beijing.
10. Cleveland, R.B. and Cleveland, W.S. (1990) STL: A Seasonal-Trend Decomposition Procedure Based on Loess. Journal of Official Statistics, 6, 3-33.
11. China Civil Aviation Administration Development Planning Department (2014) From the Statistical View of Civil Aviation. China Civil Aviation Publishing House, Beijing.
12. Liu, J.P. and Wang, Y.Q. (2015) The Historical Evolution and New Development Trend of Seasonal Adjustment Method. Statistical Research, 32, 91-98.
13. Song, Y.L. (2014) Some Properties on Extended Bayes Information Criterion. Central China Normal University, Wuhan.
14. Wu, X.Z. (2013) Statistical Method of Complex Data. China Renmin University Press, Beijing.

CHAPTER 11

Present Assessment of Public Traffic System Based on GIS in Kitakyushu

Kazuyuki Watari[1,2], Weijun Gao[1]

[1]Faculty of Environmental Engineering, The University of Kitakyushu, Kitakyushu, Japan

[2]Geo Cluster Co. LTD, Japan

ABSTRACT

With increasing of serious environment problems and the coming of the aging society, new traffic systems should be urgently rebuilt. It is necessary to evaluate the traffic service level by utilizing new technique and tool. This paper takes Kitakyushu city as case study to carry out an evaluation of urban traffic system. The concept of accessibility and mobility is used to

Citation: Watari, K. and Gao, W. (2014), "Present Assessment of Public Traffic System Based on GIS in Kitakyushu". *Open Journal of Civil Engineering*, **4**, 397-405. doi: 10.4236/ojce.2014.44034.

evaluate the present condition of existing public traffic system based on GIS technology, and then establishment of new route is discussed according to the evaluation results. Additionally, this paper established GIS database and illustrated the relationship between the public traffic and population density. The regions without enough existing traffic were identified. Moreover, the essential terms and considerations were put forward to simulate a new route of public traffic line.

Keywords: Urban Public Traffic, GIS Technology, Kitakyushu, Japan

INTRODUCTION

Advanced urban traffic system needs to be adjusted due to urgent demand of national/city's government. Therefore, it is important for urban traffic management department to put forward an improvement measure. Meanwhile, with increasing of serious environment problems (energy circumstances and urban traffic) and the coming of the aging society, new traffic systems should be urgently rebuilt. Moreover, the urban traffic plan needs to be further examined. So, it is necessary to evaluate the traffic service level by utilizing new technique and tool.

There are many researches with the traffic system. Thong [1] , developed a GIS database with consideration of efficient data input and retrieval process in a way that is most useful to transportation planners for network travel demand projection and evaluation purposes. Arampatzis [2] presented a decision support system (DSS) integrated in a geographical information system (GIS) for the analysis and evaluation of different transport policies. Suzanne Mavoa [3] calculated a public transit and walking accessibility index for Auckland, New Zealand. SUN [4] investigated the inherent principles of spatial-temporal evolution of traffic congestion based on Shenzhen floating car data (FCD) and geo-simulation platform. Charlotte Wahl [5] studied public participation in Swedish municipal traffic planning with the focus onto what extent and how municipalities and consultants interact with participants in traffic-planning processes. Lee [6] proposed a wiki-like collaborative real-time traffic information collection, fusion and sharing framework, which includes user-centric traffic event reacting mechanism, and auto- matic agent-centric traffic information aggregating scheme.

This paper takes Kitakyushu city as a case study to carry out an evaluation of urban traffic system. The concept of accessibility and mobility is used to evaluate the present condition of existing public traffic system based on GIS spatial technology, and then establishment of new urban traffic route is discussed according to the evaluation results. Then, the paper establishes the region item and discusses new routes and their profit.

STUDY AREA

Kitakyushu (33°52'N, 130°49'E) locates in Fukuoka prefecture, Kyushu, Japan. Its total area is 486.81 km^2, population is 976,800 in 2013 [7] , and population density is 2040.80 per km^2 (Figure 1). In the view of topographical features, topographic relief in Kitakyushu is bigger, sea extent elevation reaches to above 900 meters, mid- mountain and low-mountain accounting for 35% of total region area is main topographic types. Kitakyushu borders on the main islands of Japan across Kanmon Strait and is adjacent to several other

Asian countries, and particularly it is conveniently located at the straight line connecting two largest cities, Tokyo in Japan and Shanghai in China. Meanwhile, the eastern and northern Kitakyushu has long coast line. It is one of the most active areas in Kitakyushu and enjoys the fastest economic development. In the 2013, GDP is JPY 3,523,636 million, travel people is 1220 ten thousand.

INVESTIGATION OF PUBLIC TRAFFIC SERVICE

Firstly, this study extracts blank and inconvenient areas of public traffic service in Kitakyushu as research ob- jectives on the basis of GIS database.

Secondly, operating route and attained profit in these areas are evaluated.

Finally, feasibility of the new public traffic system is discussed according to the evaluation results.

Case analysis flows based on GIS technology are seen in Figure 2. The six cases are seen in Figure 3.

The cases area assumed as follow to evaluate the traffic blank area:

The inconvenient where operation new route to the bus stop is lower than 1 are extracted. The street where one-way lines are lower than 16 from am 6 to pm 10 is considered. Seen from Figure 4, the traffic blank area is outside the buffer, and the traffic inconvenient area is inside the buffer. The repeated points of each data are deleted by manual operation.

The center of each terminal in the six cases is set the center of a circle, a distance is taken as radius, the resulting circle is considered to blank and inconvenient areas. The inconvenient area increases according to the extraction data (500, 300 and 200 m away from bus stop). According to investigation results, population without access to traffic will increase with the distance away from bus stop.

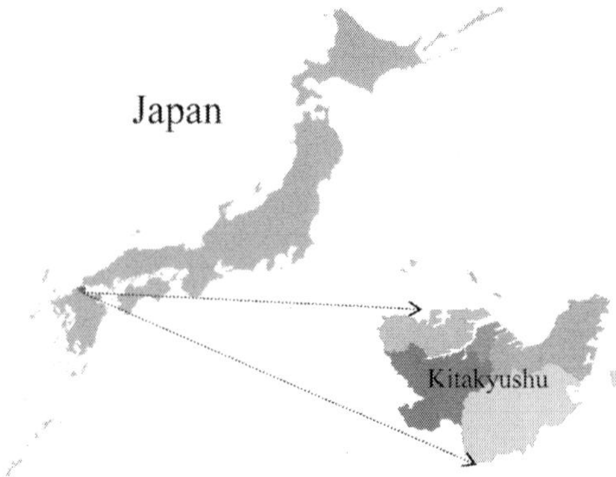

Figure 1. Study area.

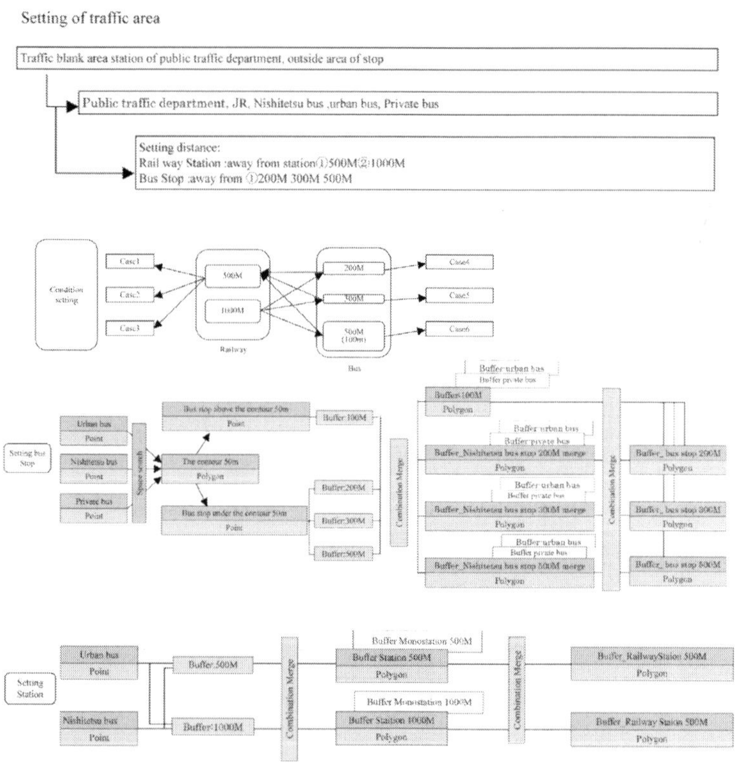

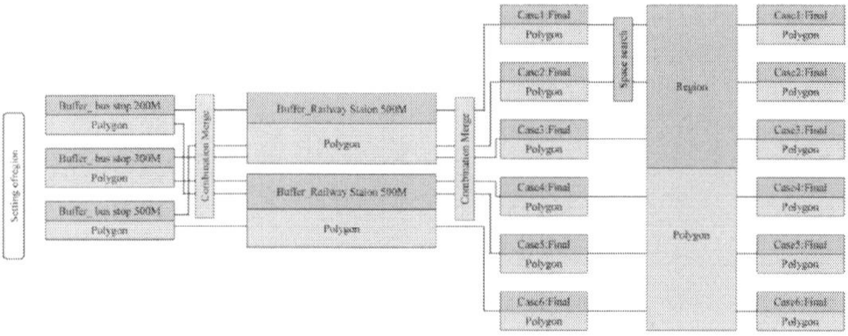

Figure 2. Case analysis flow based on GIS technology. ① 500 m away from railway and 200 m away from bus stop; ② 500 m away from railway and 300 m away from bus stop; ③ 500 m away from railway and 500 m away from bus stop; ④ 1000 m away from railway and 200 m away from bus stop; ⑤ 1000 m away from railway and 300 m away from bus stop; ⑥ 1000 m away from railway and 500 m away from bus stop.

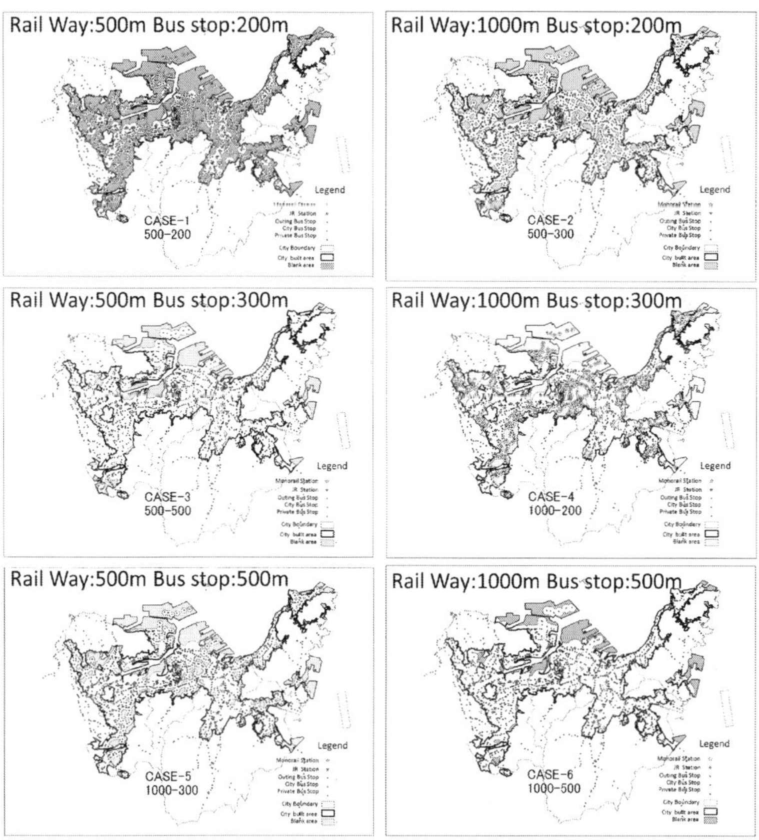

Figure 3. Case analysis.

ESTABLISHMENT OF REGION ITEM

According to the extracted elementary school, this paper selects some school areas where there are more than 2000 people and more than 500 elders as research objectives, and then neighborhoods with 1 km radius around the research objective are considered as the same areas to carry out case studies. The case studies contain CASE2 and CASE5. The number of the extraction was as follows: CASE2: 26 districts (Figure 5); CASE5: 11 districts (Figure 6). In order to discuss new route setting of target district, the following conditions need to be proposed. In CASE 2, inconvenient area is extracted to take as research objective. In the adjacent circumstances, if an area is longed to other elementary schools, but the area is situated in the 1 kilometer radius around the research objective, we think it as the same region. The requested area of government is taken as a research object.

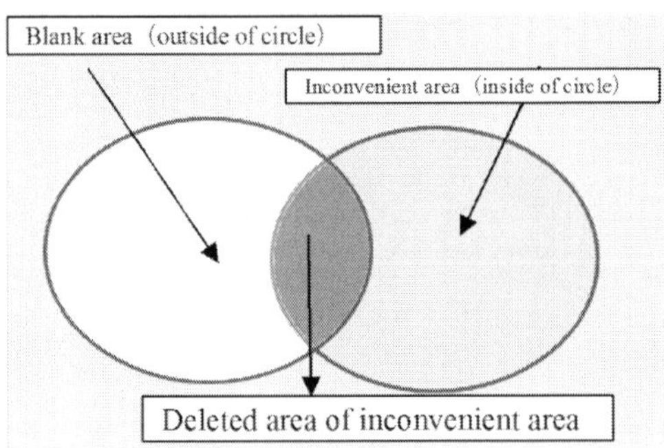

Figure 4. Blank area and inconvenient area.

Figure 5. CASE2 26 districts.

Figure 6. CASE5 11 districts.

RESULT AND DISCUSSION

Discussion of New Route

At present, bus users are inclined to decreasing. There are two reasons, on the one hand, traffic jam of the highway becomes more and more serious because of the concentration of the increasing cars; on the other hand, timing system of bus is bad. Meanwhile, there are many residents in the suburbs area. However, development balance of public bus is not adjusted, which result in that buses do not pass through the suburbs area. Until now, there aren't bus routes in the many places. Therefore, it is necessary to set up the community bus routes in the inconvenient area of public traffic. The system with short distance, small quantity, frequent transportation and the circulation type should be firstly discussed to offer comfortable mobility for every people in the narrow roads of the public transport blank (inconvenient) area (Figure 7).

It is easy to go to the urban center by operating the route linking with busy streets, the main facilities and the traffic blank zones. In addition, circulation bus of the one way will be operated; it only takes 30 - 60 minutes to run a round. So the number of bus should be increased. Since the larger route bus doesn't pass the narrow road of the residential area, if minibus is applied, the bus stops become nearer.

If the mobility of the public blank (inconvenience) area is improved, communication of local people will be activated; community will be brought up; increasing of elder's outgoing opportunities will be promoted; activation of the social participation will be supported. It is anticipated to improve the access to central town and to activate the people's communication. Importantly, the system will help the urban traffic away from seriously depending on the private cars.

Figure 7. District record.

Discussion of Profit

There are many irregular areas and ramps in the suburb; on the other hand, streets in the resident area are adjusted in the process of management. The jamming in some parts of arterial streets will make it difficult for bus to run on the dot. Even in the central area, landform and street structure still limit the convenience of traffic. Due to the willing of traveling, the amount of traveling elders has been increasing. Among the traffic ways, walking and taking bus occupy a large percent, especially, bus is usually applied in everyday life. However, some complaints are arisen against the problems when taking bus, for example, climbing up and down, receiving the message, distance to the station, bus intermediate, noisy and services.

The target people of the system are those elders who reside in the traffic-lagging areas. They use the bus in many occasions of everyday life, such as, shopping, going to hospital, and so on.

The system covers the traffic- lagging areas, areas with high elder percent and central town. The route will be regulated to single-direction circulation through the narrow streets in the period of 30 - 60 minutes.

The system will provide the people with amenity and mobility; moreover, some valuable suggestions will be zealously discussed and be introduced into the system. The distance of two bus stations is set to 200 m; additionally, waiting places are constructed under the considerations of accumulating snow. Especially, the existing establishment will be neatly made the best use.

Here about the calculation of the profit.

We assumed enterprise (investment) cost as $C_{enterprise}$ and the value of enterprise cost based on the data from the bus company.

We estimated the number of users P for the target area according to the population and age by

P = population of area × a_p (person)

Here, a_p = the use rate of people at the kind of age (based on the data from the bus company).

The income of the bus area C_{gain} can be predicted by

$$C_{gain} = sum\ (g \times P)(JPY)$$

Here g is the income for one bus line including financial aid and advertisement support (JPY/Person).

If the income of the bus area C_{gain} is larger than the $C_{enterprise}$, the project gets the profit and we say this line is "OK", otherwise we say the line is "OUT".

All the simulation was calculated in the GIS platform.

Discussion of New Traffic Service

Service time: 5 days one week, 8 hours one day, 250 days one year. Eight percent of elders (older than 65 ages) of target area will use the new traffic service, according to the data statistic of Bus Company.Charge of standard service: 23,400 Yen; Charge of subvention: 20,500 Yen; Charge of advertisement sponsorship: 20,100 - 16,500 Yen. According to the assumed conditions, profit of the new route in 2015 is simulated based on GIS technology (Table 1). "Out" means no profit under the designated condition. Therefore, in Iwai- machi area, there is no profit even with financial aid or advertisement support due to the lack of population. Profit of the 10th route under the assumed conditions is estimated by using amount of 2nd route by the 2015 will be increased because of increasing of elder's percent, which indicates that 12th route will obtain more profit. If charge of other routes can be sponsored by the advertisement agencies, possibility of getting profit will be en- sured (Figure 7).

CONCLUSIONS

In this paper, we try to present a method to verify the traffic blank and inconvenient area in Kitakyushu city.The six cases were selected to evaluate the traffic system. Based on the data of registered permanent residence, timetables of railway and road, urban management information of Kitakyushu, this study established the GIS database and illustrated the relationship between the public traffic departments and population divisions. Addi- tionally, the regions without enough existing traffic establishments were identified. Moreover, the essential terms and considerations were put forward to simulate the verifying methods for new routes of public traffic departments. However, recently price of fuels has remarkably increased, which led to the considerations again and again on arranging the outlay. We selected 22 areas to predict the profit by introducing a new bus line and gave a judgment to establish a new bus line.

In the future, we will add some GIS databases concerning about the buildings and circulating destinations, and will investigate the relations among these data; meanwhile, it is necessary to emphasize the influencing factors in the research method of new routes.

Table 1. Profit of new route in 2015.

2015		Running hour 8/ week 5 days	Standard case	Financial aid case	Advertisement support case									
No.	Element school name	250 days			100,000	200,000	300,000	400,000	500,000	600,000	700,000	800,000	900,000	1,000,000
		200JPY	23,400	20,500	20,100	19,700	19,300	18,900	18,500	18,100	17,700	17,300	16,900	16,500
1	Ibori	52,614	OK	OK	OK	OK	OK	OK	OK	OK	OK	OK	OK	OK
2	Nakai	27,408	OK	OK	OK	OK	OK	OK	OK	OK	OK	OK	OK	OK
3	Izumi-dai	27,958	OK	OK	OK	OK	OK	OK	OK	OK	OK	OK	OK	OK
4	Min-amigoka	20,551	OUT	OK	OK	OK	OK	OK	OK	OK	OK	OK	OK	OK
5	Tomino	19,133	OUT	OUT	OUT	OUT	OUT	OK	OK	OK	OK	OK	OK	OK
6	Kuzu-hara	31,365	OK	OK	OK	OK	OK	OK	OK	OK	OK	OK	OK	OK
7	Yukawa	24,528	OK	OK	OK	OK	OK	OK	OK	OK	OK	OK	OK	OK
8	Kikuga-oka	30,602	OK	OK	OK	OK	OK	OK	OK	OK	OK	OK	OK	OK

9	Sone	34,461	OK	OK	OK	OK	OK	OK	OK	OK	OK	OK	OK	OK
10	Maki-yama	56,668	OK	OK	OK	OK	OK	OK	OK	OK	OK	OK	OK	OK
11	Ookura	29,034	OK	OK	OK	OK	OK	OK	OK	OK	OK	OK	OK	OK
12	Iwaima-chi	15,621	OUT	OUT	OUT	OUT	OUT	OUT	OUT	OUT	OUT	OUT	OUT	OUT
13	Kuro-hata	21,750	OUT	OK	OK	OK	OK	OK	OK	OK	OK	OK	OK	OK
14	Kamiko-jyaku	48,012	OK	OK	OK	OK	OK	OK	OK	OK	OK	OK	OK	OK
15	Chiyo	43,034	OK	OK	OK	OK	OK	OK	OK	OK	OK	OK	OK	OK
16	Hanao	24,806	OK	OK	OK	OK	OK	OK	OK	OK	OK	OK	OK	OK
17	Touno	17,705	OUT	OUT	OUT	OUT	OUT	OUT	OUT	OUT	OK	OK	OK	OK
18	Sarakura	4,660	OUT	OUT	OUT	OUT	OUT	OUT	OUT	OUT	OUT	OUT	OUT	OUT
19	Katsuki	15,467	OUT	OUT	OUT	OUT	OUT	OUT	OUT	OUT	OUT	OUT	OUT	OUT
20	Matsug-aoka	8522	OUT	OUT	OUT	OUT	OUT	OUT	OUT	OUT	OUT	OUT	OUT	OUT
21	Hagiga-oka	18,442	OUT	OUT	OUT	OUT	OUT	OUT	OUT	OK	OK	OK	OK	OK
22	Fukama-chi	19,804	OUT	OUT	OUT	OK	OK	OK	OK	OK	OK	OK	OK	OK

ACKNOWLEDGEMENTS

Kazuyuki Watari, Weijun Gao This research was based on data from the project of Kitakyushu Transportation Bureau. We would like to ex- press our thanks to Kitakyushu city.

REFERENCES

1. Thong, C.M. and Wong, W.G. (1997) Using GIS to Design a Traffic Information Database for Urban Transport Plan- ning. Computers, Environment and Urban Systems, 21, 425-443. http://dx.doi.org/10.1016/S0198-9715(98)00003-9
2. Arampatzis, G., Kiranoudis, C.T., Scaloubacas, P. and Assimacopoulos, D. (2004) A GIS-Based Decision Support System for Planning Urban Transportation Policies. European Journal of Operational Research, 152, 465-475.
3. Mavoa, S., Witten, K., McCreanor, T. and O'Sullivan, D. (2012) GIS Based Destination Accessibility via Public Transit and Walking in Auckland, New Zealand. Journal of Transport Geography, 20, 15-22. http://dx.doi.org/10.1016/j.jtrangeo.2011.10.001
4. Sun, J. (Daniel), Liu, Q. and Peng, Z.R. (2011) Research and Analysis on Causality and Spatial-Temporal Evolution of Urban Traffic Congestions—A Case Study on Shenzhen of China. Journal of Transportation Systems Engineering and Information Technology, 11, 86-93.
5. Wahl, C. (2013) Swedish Municipalities and Public Participation in the Traffic Planning Process—Where Do We Stand? Transportation Research Part A: Policy and Practice, 50, 105-112. http://dx.doi.org/10.1016/j.tra.2013.01.012
6. Lee, W.-H., Tseng, S.-S. and Shieh, W.-Y. (2010) Collaborative Real-Time Traffic Information Generation and Sharing Framework for the Intelligent Transportation System. Information Sciences, 180, 62-70. http://dx.doi.org/10.1016/j.ins.2009.09.004
7. Kitakyushu City (2014) Population. http://www.city.kitakyushu.lg.jp/soumu/file_0373.html

CHAPTER

12

Development of a GIS Based Public Transport Composite Social Need Index in Jeddah

Mohammed Aljoufie

Department of Urban and Regional Planning, Faculty of Environmental Design, King Abdulaziz University, Jeddah, Saudi Arabia

ABSTRACT

Provision of public transport is critical for all social groups with high need. The provision of effective public transport system in Jeddah is at the top priorities of the national, regional and local plans. This paper attempts to develop a public transport composite social need index in Jeddah using GIS. Population density, car ownership, family size and income level indicators were considered for identifying the social need in Jeddah. Results indicate

Citation: Aljoufie, M. (2016), "Development of a GIS Based Public Transport Composite Social Need Index in Jeddah". *Journal of Geographic Information System*, **8**, 470-479. doi: 10.4236/jgis.2016.84039.

that Jeddah is in high social need for public transport and the provision of an efficient public transport is crucial for the city. The composite social need index has shown both dispersed and clustered patterns of high need for public transport in Jeddah. The results of this study facilitate the identification of the high social need for public transport in Jeddah. Further studies are highly recommended to include other social need indictors in Jeddah city.

Keywords: Public Transport, Social Need, Indicators, GIS, Jeddah City

INTRODUCTION

Rapidly growing cities confront with enormous challenges. High population growth rates, increased expansion of urban areas stimulate traffic and high dependence on private vehicle, causing congestion and consequent effects, for instance air pollution, greenhouse gas emissions and economic losses [1] [2] .

Saudi cities are not an exception; main cities in Saudi Arabia experience these challenges. Jeddah, the second largest city, has undergone a rapid urban growth, land use changes and spatial expansion during the past 50 years. The population of Jeddah has rapidly grown from about 147,900 in 1964 to more than 3.5 million nowadays. In line with that, the urban area has extended from about 18,315 hectare in 1964 to about 60,000 hectare nowadays [2] . In addition to that, transportation infrastructure has expanded dramatically from about 101 km in 1964 to more than 850 km nowadays.

These changes have catalyzed notable travel pattern effects. Urban sprawl, lack of integrated planning frame- work and dramatic land use changes have caused a significant impact on the modal split [2] . The share of daily non-car trips has decreased dramatically. Trips with public transportation have decreased from about 19% of all travel modes in 1970 to about 2.3% in 2007 [2] . Moreover, the share of other modes, such as cycling and pedestrian trips, has declined notably from about 31% in 1970 to about 4.6% in 2007 [2] . Conversely, the percentage of daily trips by car has increased dramatically from about 50% in 1970 to about 96% in 2012. This has coincided with poor public transport coverage and service in Jeddah city.

This has stimulated high levels of mobility all over the city wherein congestion is widespread on Jeddah's streets in the morning and evening peak periods and in non-peak hours [3] [4] . The average speed on the highways and the main roads is lessening, and traffic safety is decreasing, while transportation emissions and average trip durations are increasing [2] [4] .

To confront theses challenges, encouragement of effective public transport is vital. In fact, public transport is becoming an alternative to the private vehicle and highly used by different socioeconomic groups, given the high costs of individual transport, impact on environment, and continues increase of congestion in cities [5] .

The provision of effective public transport is at the top priorities of the national, regional and local plans. It is a major priority in Jeddah city; thus understanding of the social need for public transport has received a little attention.

Geographic Information Systems (GIS) represent powerful tools for urban management and spatial analysis. GIS functionalities support the examination of urban spatial analysis, trends, and urban systems interaction [6] . Developed methods of spatial analysis have shown rich opportunities for urban analysis [7] . This paper attempts to develop a public transport composite social need index in Jeddah using GIS.

METHODOLOGY

Study Area

Jeddah is the second largest city in Saudi Arabia. Jeddah located in the western region of Saudi Arabia on the Red Sea eastern coast (Figure 1).

The rapid rates of urban growth over the last 5 decades have caused several challenges for Jeddah city. The travel pattern has been affected accordingly and this has catalyzed the car dependency wherein Jeddah people use private vehicle for about 96% of daily trips [8] . This has coincided with deficient public transport system [9] . Accordingly the provision of efficient public transport system is critical. However, to an efficient public transport, understating the social need for public transport is vital. This issue has received a little attention. Thus, this study attempts to develop a composite social need index for Jeddah using GIS. The extent of urban districts boundaries is considered in this study.

Public Transport Social Need Composite Index

In essence, various factors influence the need for public transport. Murray and Davis [10] developed an index to determine the public transport need based on attribute indicators of spatial units. Three indicators were used, i.e.: the family income, unemployment rates, and average family size. A weighting approach that combines these indicators was used to evaluate the public transport needs for each zone within the study area.

Currie [11] [12] proposed a transport needs index at district level. This index encompassed indicators for transport needs, namely: adults without cars, persons aged over 60 years, persons on a disability pension, low income households, adults not in the labour force, students and persons 5 to 9 years.

Yao [13] used a composite need index to measure the relative potential demand for public transport in a spatial unit. Three groups of factors for public transit ridership for commuting trips in Atlanta were identified. The first group was the land use characteristics which include: population density, employment rate (percentage of people who have jobs), job density (total jobs in the TAZ divided by the area of the TAZ), average number of jobholders per household, and percentage of home workers. The second group was the socioeconomic characteristics which encompass: income and car ownership. The third group was the network structure which includes the density of public transit stops in the TAZ.

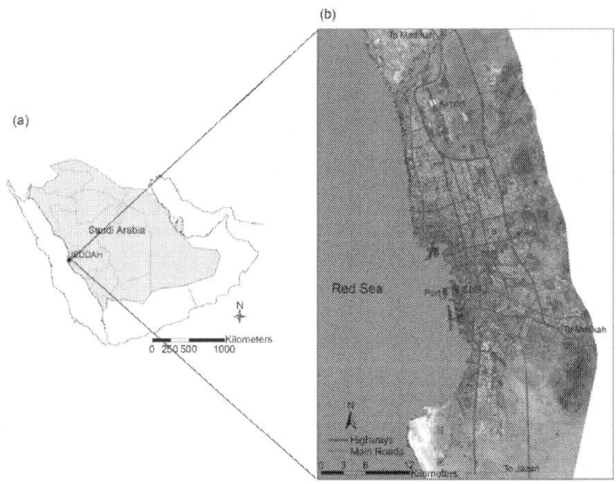

Figure 1. (a) Jeddah city location; (b) Jeddah urban boundaries.

In view of that four social indicators were considered. Table 1 depicts the selected indicators which were based on a review of available socioeconomic data and related to public transport need. Population density, car ownership, family size and income level indicators at district level were considered.

A consistent weighting approach that evaluates the public transport social needs for each district within the study area was used. Each indicator was classified into three weights as followings:

- High: high need for public transport,
- Medium: medium (moderate) need for public transport, and
- Low: low need for public transport.

An absolute need scores from 3 to 1 (High = 3, Medium = 2 and Low = 1) were used in the weighting to standardize the value of indicators. Weight 3 represents districts with high social need for public transport while weight 1 represents districts with low social need for public transport. The weightings of each indictor into the three classes were based on the available data and sources [14] [15] .

Subsequently, a composite social need index was developed. It was generated by adding the scores for each indicator together as followings:

$$CSNI = \sum PdI + CI + FsI + InI$$

where CSNI is the composite social need index, PdI is the standardized population density indicator, CI is the standardized car ownership indicator, FsI is the standardized family size indicator and InI is the standardized income indicator.

The composite social need index was standardized between 3 and 12 for all districts in the analysis in order to obtain need scores. Value 3 represents districts with low social need for public transport while value 12 represents districts with low social need for public transport. To quantify the composite social need, the results were then divided into three groups as followings:

- High: from 8 to 12.
- Medium: from 7 to 8.
- Low: from 3 to 6.

Table 1. Selected indicators and weight.

Indicator	Description	Weight
Population density	Person/hectare per district	3) High: >150 person per hectare 2) Medium: 75 - 150 person per hectare 1) Low: <75 person per hect-are
Car owner-ship	Number of car per 100 persons per district	3) High: <20 car/100 persons 2) Medium: 20 - 30 car/100 persons 1) Low: >30 car/100 persons

Family size	Average family size per district	3) High: >5 person 2) Medium: 3 - 5 person 1) Low: ≤3 person
Income level	Average family income per district	3) High: ≤5000 SAR 2) Medium: 5000 - 15,000 SAR 1) Low: >15,000 SAR

Data Collection and Preparation

Different data sources were used to quantify the public transport social need indicators in order to calculate the composite social need index. Table 2 indicate the data sources of the considered indicators.

For population density census data of 2007 were used. In conjunction with that, Jeddah urban observatory data of 2009 were considered for car ownership, car ownership, family size and income level indicators. However, the urban observatory data of 2009 were available for 58 districts out of 117 districts (the total of Jeddah urban districts). Thus, a questionnaire designed to collect data o the remaining districts.

The questionnaire has included three questions (closed-ended multiple-choice questions) to define data of car ownership, family size and income level indicators.

The questionnaire was simplified to take about 5 minutes in maximum to be completed. This timing was vital in order to raise the respondent rate. The survey was performed during the period from 10th March 2015 to 10th May 2016. The survey was conducted at the main attractions of each district (has no data) including shopping centres and main commercial arterials. Moreover, the questionnaire was posted online also using different social network sites. The respondent rate was high and the completed questionnaires were considered in the analysis.

RESULTS AND DISCUSSION

The indicators of public transport social need in Jeddah are determined by district. Table 3 and Figures 2-6 show the results of the spatial analysis of the considered public transport social need indicators and the developed composite social need indicators.

Population density indicator depicts high transport need in different districts throughout Jeddah, Figure 2. It is noted that low need is dominant with this indicator as compared with high and medium need which reflected by the number of districts as depicted in Figure 2 and Table 3. However, high and medium need districts exhibit majority of the population. It is noted that high and medium need districts concentrate at the city centre and drawn areas. Conversely, low need districts distributed all over Jeddah urban peripheries.

Figure 3 show the spatial distribution of the absolute scores of car ownership indicator. The districts which show high public transport need are concentrated in the city centre. The population with high and medium need are foremost under this indicator as compared with population with low need, as indicated in Table 3. However, medium need districts are dominant which reflected by the number of districts as depicted in Figure 3 and Table 3. Medium need districts distributed all over the Jeddah urban peripheries and in some districts drawn to the city centre.

Family size indicator depicts high transport need in different districts throughout Jeddah, Figure 4. It is noted that high need is dominant with this indicator as compared with medium and low need which reflected by the number of districts as depicted in Figure 4 and Table 3. It is also noted that high need districts exhibit majority of the population. It is noted that high and medium need districts concentrate at the city centre and drawn areas and disperse at north and south urban peripheries. Conversely, low need districts distributed all over Jeddah urban peripheries in a disperse pattern at the north and the east and concentrate at the south of Jeddah.

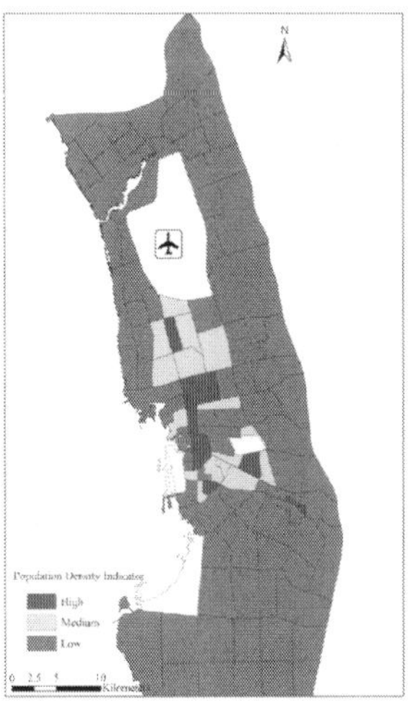

Figure 2. Population density indicator.

Table 2. Selected indicators and weight.

Indicator	Description	Source
Population density	Person/hectare per district	Census 2007
Car ownership	Number of car per 100 persons per district	-Jeddah Urban Observatory 2009* -Survey
Family size	Average family size per district	-Jeddah Urban Observatory 2009* -Survey
Income level	Average family income per district	-Jeddah Urban Observatory 2009* -Survey

*Jeddah Urban Observatory (2009). Districts GIS Indicators, Jeddah Urban Observatory, Jeddah Municipality.

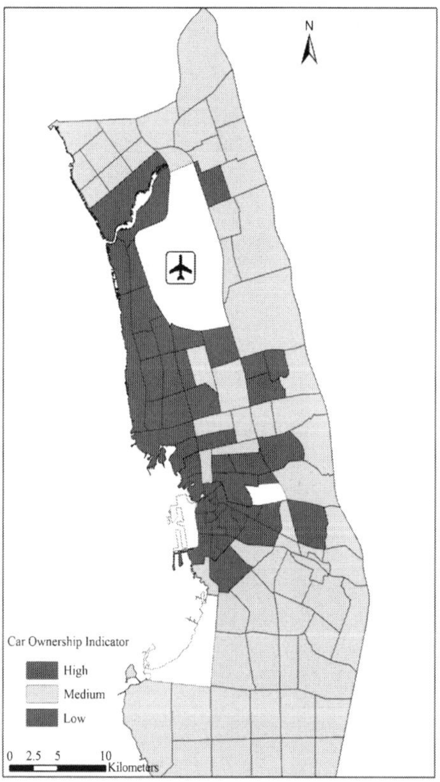

Figure 3. Car ownership indicator.

Table 3. No. of districts and population by transport need for each indicator.

Need	Population density indicators		Car ownership indicator		Family size indicator		Income level indicator		Composite social need index	
	Dist.	Pop.	Dist.	Pop.	Dist.	Pop.	Dist.	Pop.	Dist.	Pop.
High	14	876,831	23	834,142	49	1,598,123	25	615,505	27	1,524,833
Medium	16	1,181,026	62	945,999	25	1,218,341	69	1,779,761	35	892,056
Low	83	781,379	30	1,059,095	39	22,772	19	443,970	51	422,347

Figure 5 show the spatial distribution of the absolute scores of income level indicator. The districts which show the high public transport need are concentrated in the city centre and slightly disperse at the east and the south of Jeddah. The population with high and medium need are foremost under this indicator as compared with population with low need, as indicated in Table 3. However, medium need districts are dominant which reflected by the number of districts as depicted in Figure 5 and Table 3. Medium need districts distributed all over the Jeddah urban peripheries and in some districts drawn to the city centre.

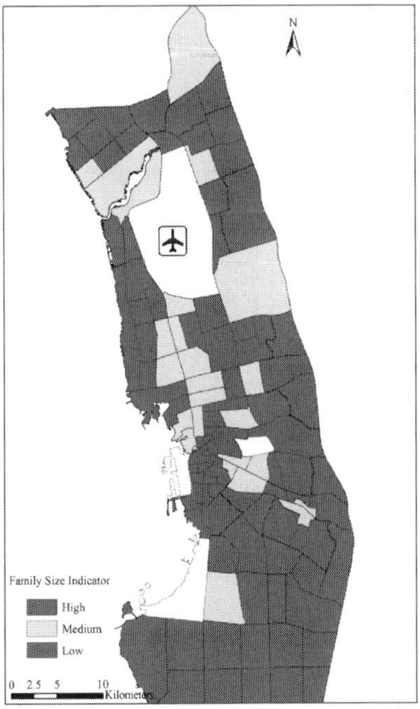

Figure 4. Family size indicator.

It is noted that car ownership indicator and income level indicator exhibit a comparable spatial distribution of public transport need, as shown in Figure 3 and Figure 5. However, the population of these indicators differs. Conversely, population density indicator and family size indicator exhibit a different spatial distribution pattern of public transport need. These indicators also show a different population of high, medium and low needs.

The result of composite social need index, which combines all the considered four indicators, show both dispersed and clustered patterns of high need for public transport in Jeddah. It is found that districts with high social need are concentrated and clustered in the city centre, while some districts are found at the north of the city centre, Figure 6. Medium need also exhibit the clustered and a dispersed pattern of districts at area drawn to the city centres and at the north, east and south of Jeddah city. It is noteworthy that high and medium need districts are dominant and spatially connected, Figure 6. Notably, about 53% of Jeddah population are in high social

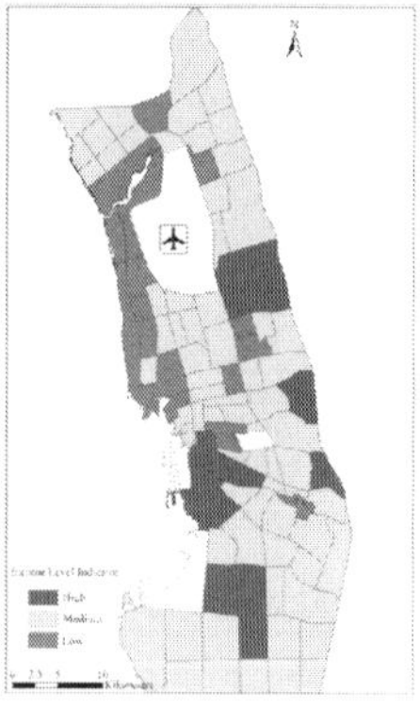

Figure 5. Income level indicator.

need for public transport, about 32% are in medium social need for public transport, and about 15% of Jeddah population are in low social need for public transport, Table 2. This indicates that Jeddah is in high social need for public transport and the provision of an efficient public transport is crucial for the city.

Public transport is important for social inclusion, for providing access to reach activities and services such as work, education, health, shopping and social-recreational activities [16] . The provision of public transport depends on demand which is based on the population need. This paper has developed a composite social need index for Jeddah using GIS. GIS has shown the capability of identifying, visualizing and calculating the social need for public transport. This index facilitates the identification of social need for public transport in Jeddah and assists transportation planners for high social inclusion in planning the routes of public transport system.

CONCLUSION

This paper has developed a composite social need index for public transport in Jeddah using GIS. Population density, car ownership, family size and income level indicators were considered for identifying the social need in Jeddah.

Figure 6. Composite social need index.

The composite social need index has shown both dispersed and clustered patterns of high need for public transport in Jeddah. Districts with high need are concentrated and clustered in the city centre, while some districts are dispersed at the north of the city centre. This index reveals that Jeddah is in high social need for public transport and the provision of an efficient public transport is crucial for the city. The results of this study facilitate the identification of potential demand for public transport in Jeddah and assist transportation planners for high social inclusion in planning the routes of public transport system. GIS has shown the capability of identifying, visualizing and calculating the social need for public transport. Although this study has provided valuable information, it must be noted that available data don't depict all the social need in Jeddah. Thus, further studies with the same methodology are highly recommended to include other social need indictors in Jeddah city.

REFERENCES

1. Zhao, P. (2010) Sustainable Urban Expansion and Transportation in a Growing Mega City: Consequences of Urban Sprawl for Mobility on the Urban Fringe of Beijing. Habitat International, 34, 236-243. http://dx.doi.org/10.1016/j.habitatint.2009.09.008
2. Aljoufie, M., Zuidgeest, M.H.P., Brussel, M.J.G. and van Maarseveen, M.F.A.M. (2013) Spatial-Temporal Analysis of Urban Growth and Transportation in Jeddah City, Saudi Arabia. Cities, 31, 57-68. http://dx.doi.org/10.1016/j.cities.2012.04.008 [Citation Time(s):6]
3. IBI, Group, Jeddah Public Transport Study, Ministry of Transport, Saudi Arabia, 2007.
4. Jeddah Municipality, Jeddah Strategic Plan, Saudi Arabia: Jeddah Municipality, 2009.
5. Tribby, C.P. and Zandbergen, P.A. (2012) High-Resolution Spatio-Temporal Modeling of Public Transit Accessibility. Applied Geography, 34, 345-355. http://dx.doi.org/10.1016/j.apgeog.2011.12.008
6. Benenson, I. and Torrens, P.M. (2004) Geosimulation: Automata-Based Modeling of Urban Phenomena. John Wiley and Sons Ltd, Chichester. http://dx.doi.org/10.1002/0470020997
7. Paez, A. and Scott, D. (2004) Spatial Statistics for Urban Analysis: A Review of Techniques with Examples. GeoJournal, 61, 53-67. http://dx.doi.org/10.1007/s10708-005-0877-5
8. AECOM (2012) Jeddah Spatial Framework Plan in Association with the Jeddah Stormwater Drainage Program. Emirate of Makkah, Saudi Arabia.
9. Aljoufie, M. (2016) Exploring the Determinants of Public Transport System Planning in Car-Dependent Cities. Procedia-Social and Behavioral Sciences, 216, 535-544. http://dx.doi.org/10.1016/j.sbspro.2015.12.013
10. Murray, A.T. and Davis, R. (2001) Equity in Regional Service Provision. Journal of Regional Science, 41, 577-600. http://dx.doi.org/10.1111/0022-4146.00233
11. Currie, G. (2004) Gap Analysis of Public Transport Needs: Measuring Spatial Distribution of Public Transport Needs and Identifying Gaps in the Quality of Public Transport Provision. Transportation Research Record: Journal of the Transportation Research Board, 1895, 137-146. http://dx.doi.org/10.3141/1895-18

12. Currie, G. (2010) Quantifying Spatial Gaps in Public Transport Supply Based on Social Needs. Journal of Transport Geography, 18, 31-41. http://dx.doi.org/10.1016/j.jtrangeo.2008.12.002
13. Yao, X. (2007) Where Are Public Transit Needed—Examining Potential Demand for Public Transit for Commuting Trips. Computers, Environment and Urban Systems, 31, 535-550. http://dx.doi.org/10.1016/j.compenvurbsys.2007.08.005
14. Jeddah Urban Observatory, Districts GIS Indicators, Jeddah Urban Observatory, Jeddah Municipality, 2009.
15. Jeddah Municipality, Jeddah structure plan, Saudi Arabia: Jeddah Municipality, 2004.
16. Lucas, K. (2010) Transport and Social Exclusion: Where Are We Now? Paper Presented to the 12th World Conference on Transport Research, Lisbon, 11-15 July 2010.

CHAPTER 13

Evaluation of Past Investment in Urban Public Transportation

Nir Sharav, Yoram Shiftan

Transportation Research Institute, Technion, Haifa, Israel

ABSTRACT

Mobility is a major concern in cities all over the world. Population and density increase makes urban mobility more complex to plan and budget. Decision makers have to choose which transport projects to build and how much budget to allocate to different types of investments. Cities vary by many characteristics (size, density, etc.), and different cities have adopted different transport solutions. Some cities invested more in road networks while others invested more in public transport (PT) networks. Questions regarding the amount invested in

Citation: Sharav, N. and Shiftan, Y. (2017), "Evaluation of Past Investment in Urban Public Transportation". *Theoretical Economics Letters*, **7**, 543-561. doi: 10.4236/tel.2017.73040.

urban public transport and urban roads and the right balance between these investments, taking into account the urban characteristics and the residents' preferences, has received less attention in the literature. In this research, we focused on urban public transport investments in various cities and examine the relationship between public transport and road network investments, speed, GDP, and modal split. The results showed that in developed cities, the current investment in public transport contributes to PT usage and increases PT share. Public transport reserved routes (as an indicator for PT inventory or past investments), jobs proportion in the Central Business District(CBD), and public transport supply were also found to have a positive effect on PT modal split, while motorization level was found to have a negative impact on PT usage as expected. The analysis showed that cities invested on average 7 - 8 thousand US dollars per capita in the public transport infrastructure, accounting for about 50% of the total transport budget. Cities with more developed public transport system invested about 15 thousand US dollars per capita, and allocated 65% of the budget to public transport. These cities manage to maintain the average public transport speed in the range of 30 km/h (on average a 1.3 km/h improvement in the public transport average speed for every 1000 dollar investment per capita). The investments in cities with developed public transport systems generated time benefits that covered on average 0.6 - 0.7 of the investment. Some cities have B/C ratios higher than 1.0, demonstrating that the time benefits predicted by the model covered the investment.

Keywords: City Level Modal Split Model, Urban Public Transport Investment Model, Increase in Public Transport Speed, Public Transport Time Benefits

INTRODUCTION

Mobility is essential and a major concern in cities all over the world. Increase in population and density makes urban mobility more complex to plan and budget. Limited land and budget resources on the one hand, and growing demand and congestion on the other hand, make city transport planning crucial to enable accessibility, improve productivity and life quality and support activities and economic growth.

Cities vary by size, density, urban structure, population, employment, and socio-economic characteristics. These differences are also reflected in the transport characteristics and the mobility solutions each city has developed. Newman & Kenworthy [1] showed the correlation between urban density and transport energy consumption per capita. They showed that cities with high urban density have lower transport energy consumption. Commute pattern and modal split also differ significantly among cities. For example, in some US cities (such as Dallas, San Diego, and Columbus) more than 95% of the commuters travel alone by car, while in other cities public transport accounts for about 30% of total commuters (Boston, Chicago, San Francisco and Washington DC). Some cities such as Barcelona, Berlin, Tokyo, Osaka, Prague, Singapore, London, and Paris rely heavily on public transport, which accounts for more than 50% of all motorized trips.

The relationship between transport investments and the economy has been a subject of extensive research in both micro and macro-economic levels. The relationship between transport investment, congestion, road pricing, and transit fare and subsidy is also subject to extensive theory development and research.

The micro economic theory of transport investment was developed more than fifty years ago, when the fundamentals of welfare theory as the basis for cost benefit analysis (CBA) were introduced. Berechman [2] described the latest theory and practice evaluation of transport projects. This theory is now in com- mon use in many countries as a practical method for transport project appraisal (see the international comparison by Hayashi, Y. & H. Morisugi [3] , Mackie & Worsley [4] , and also Vickerman [5] for the United Kingdom, and Shiftan, Sharaby & Solomo [6] for Israel.

New theoretical development has also incorporated macro-economic analysis into the micro-level CBA practice in forms of agglomeration effects and other wider economic impacts [7] [8] [9] . The macro-economics of transport investments usually identify the impact of a transport investment on the gross domestic product (GDP), the labor market and the real estate market. The research in this area is very extensive and shows big variations in the results and magnitude of impact. An extensive review of different approaches and research is described in Banister & Berechman [10] and Berechman [2] .

Empirical research of urban public transport investments in cities around the world showed that cities with higher urban density, where more trips are made by public transport, cycling, or by foot, are more efficient in terms of mobility costs [11] .

Trip costs in sprawling cities (especially in North America and Oceania where more than 90% of inner city trips are made by car) are 50% more expensive than Western European cities and over 100% more costly than affluent Asian cities. Trip costs to the community are 12.5% of GDP in the USA and Canada, 8.3% in Western Europe, and 5.4% in Asian cities.

Vivier et al. [11] showed that on average, public transport consumes four times less energy per passenger km than transport by car (in Canada three times less, in Europe 3.7 times less, and in Japan ten times less.)Sprawling cities in the USA, Canada and Australia have well developed road networks and high motorization levels. Wetern European cities and affluent Asian cities invested more in public transport, hence the ratio of public transport exclusive rights-of-way km to motorway km is more than seven times the average of US and Canadian cities. Asian cities with high density are characterized by developed mobility that is less car dependent and based more on walking, cycling, and publictransport. These cities invested more in public transport with average lenghs that are four times greater than their motorway spans. Wetern European cities are more similar to affluent Asian cities in terms of investment and usage of public transport systems.

Large cities reveal a fundamental difference between private and public trans- port cost on the supply side. On the one hand, congestion causes increases in the private and public costs of road transportation, while on the other hand, higher population density and investment in public transport infrastructure lowers the cost of using public transport. The cost characteristics of urban transport make urban transport planning difficult and tricky. Investment in city roads increases private car usage and decreases public transport ridership, and thus increases congestion and total generalized costs. This is known as the Down-Thompson Paradox [12] . On the other hand, investment in urban public transport tends to be capital intensive. Berechman [2] showed that inferior transport mega-projects are often selected.

Basso & Jara Diaz [13] developed a combined private car and transit model in which travelers can choose between two modes, car or transit, based on the generalized cost they perceive. The welfare maximization is optimized by three decision parameters: congestion price, transit fare (and hence transit subsidy) and transit frequency representing the investment in transit.

Beaudoin, Farzin, & Lin [14] estimated the effect of past public transit investment on traffic congestion and showed that increases in public transit supply lead to a small overall reduction in auto traffic congestion. The elasticity of auto travel with respect to transit capacity varies from −0.02 for smaller, less densely populated regions with less-developed public transit networks, to −0.4 in the largest, most densely populated regions with extensive public transit networks.

Policy makers, city planners, and researchers each play a role in planning and developing of the city's transport system, choosing which projects should be built, and allocating budget among different types of transport investments such as roads, sidewalks, public transport, and bicycle lanes.

In most countries, an economic appraisal of the projects is made. Planners have to decide how much budget should be invested on transportation, and how this budget should be allocated between public transport projects and roads based on their plans, the availability of funds, potential projects and the benefits of the various projects.

The objective of this research is to analyze past urban public transport investments per capita in various cities and the budget allocation between roads and public transportation. We also aim to study the impact of public transport investments on modal split, speed and accessibility in terms of time savings and benefits.

Chapter 2 describes the research methodology and the city level economic model, and the cities data used in this research. Chapter 3 presents the empirical model results and analysis including the city level modal split model, public transport speed model and the benefits of public transport investments in the cities. Chapter 4 presents the main conclusions of the effects of the investments on public transport speed, modal split and time benefits. The cities included in this research are described in the Appendix.

METHODOLOGY

We developed a cost benefit economic model based on welfare theory as described in Berechman [2] . The difference is that we use some unique approach to the traditional CBA. While CBA is usually used to analyze a specific project, here we calculated time benefits from an investment policy, as set by the actual annual investment made in each city. This approach uses the same theory as the project level cost benefit analysis, but assigns the theory to the network level long term investment. The welfare theory indicates that network investments should produce long term city level benefits. The model described in this chapter aims to capture these benefits.

We used city-level macro data analysis obtained from different sources that are described later in this chapter. We only considered the direct time benefits generated by the transport investment, as the main benefit that estimates (in money terms) the accessibility improvements generated by the project. Time benefits represent a minimum base line of the benefits generated by transport investments. Other benefits such as safety and environmental influences, economic development, and agglomeration are not included in the model and require further research and modeling.

The Model

We assume a basic aggregate two mode fix demand economic model with city population of Npersons that produces n daily trips which can be described as city commuters who use either public transport or cars.

Figure 1 illustrates the effect of public transport investment in a two-mode model. The equilibrium point E_0 describes equilibrium in the city transport network with n_{pto} public transport commuters and nc_o car commuters, where $n_{pto}+n_{c0}=n$.

We now assume an annual investment in public transport II that reduces travel costs (such as time) and increases public transport attractiveness and usage. In a two-mode fix demand model, this will imply lower demand for private car trips and a change of modal split into higher usage in public transport (In reality, induced demand of new trips might be attracted to the roads and thus the congestion savings might be less then estimated in the fix demand model, however, this is compensated by not including the benefits of the new trips in the fix demand model).

Point E_1 on Figure 1 describes the new equilibrium after the investment had been made. The public transport supply shifts from S_o to S_1 , reduces public transport costs (P_{pt1}) and induces private car users to switch to public transport ($n_{pt1}-n_{pt0}$). The decreased demand for private car trips (D_{1P}) reduces car commuters to n_{c1} and travel costs to P_{c1} .

By the rule of half, the investment in public transport will increase social welfare if total benefits exceed total investment costs as described in Equation (1):

$$\underbrace{I \le \left(P_{pt0} \times n_{pt0} - P_{pt1} \times n_{pt0}\right)}_{\substack{\text{Benefit to existing PT users} \\ \text{(travel time savings)}}} + \underbrace{\frac{1}{2}\left(P_{pt0} - P_{pt1}\right)\left(n_{pt1} - n_{pt0}\right)}_{\text{Benefit to new PT users}} + \underbrace{\left(P_{c0} - P_{c1}\right) \times n_{c1}}_{\text{Benefit to car users}} \quad (1)$$

where,

I = investment in public transport infrastructure,

n = number of trips (city commuters),

With sub index 0 or 1 for before and after the investment respectively,

n_{pt0} = number of travelers by public transport, before the investment,

n_{pt1} = number of travelers by public transport, after the investment,

n_{c1} = number of travelers by car after the investment,

N = city population,

Pc = private car costs (car user's travel time cost),

Tc = car travel time,

Tpt = transittravel time,

Ppt = public transport costs (user's travel time cost).

And we assume in this model that:

$$n_{pt0} + n_{c0} = n_{pt1} + n_{c1} = n$$

As described above, we assume that all costs are time costs, and define:

VOT = value of time,

DTP = daily trips per person.

And so we get:

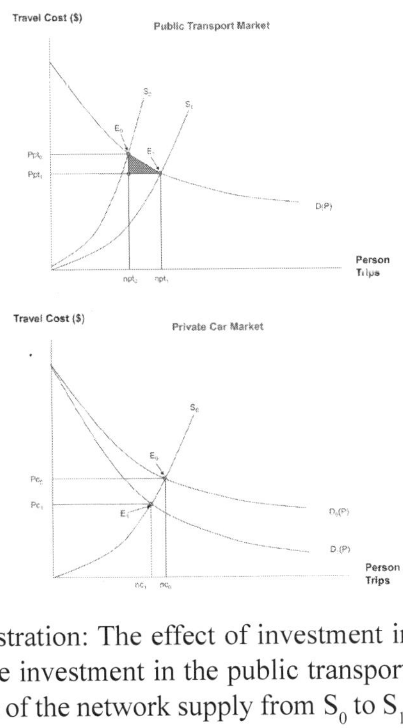

Figure 1. Model illustration: The effect of investment in public transport in a two-mode model. The investment in the public transport network is presented in the figure as a shift of the network supply from S_0 to S_1. This effct is then followed by reduction of car trip demand from D_0 to D_1. The benefits are described in Equation (1).

n=N×DTP

Define modal split (share of public transport travelers) as $MS_0 = \frac{n_{pt0}}{n}$ and divide Equation (1) by N and replace:

$$Ppt = Tpt \times VOT$$

$$Pc = Tc \times VOT$$

We get:

$$\frac{I}{N} \leq \left[\frac{1}{2} \left(MS_0 \times \Delta T_{pt} + MS_1 \times \Delta T_{pt} \right) + \left(1 - MS_1 \right) \times \Delta T_c \right] \times VOT \times DTP \tag{2}$$

Equation (2) is an interesting presentation of the standard Cost Benefit calculation.

On the left side we get the investment per capita, which can now be compared to the result on the right side that shows that the benefits can be represented as:

- The change in travel time in public transport that the investment has caused.
- The change in PT share (increase in public transport trips switching from car trips).
- The change in road network travel time caused by the reduction in car trips and congestion.
- The value of time and average trips per person (per day).

Equation (2) further implies that larger cities should have higher investment per capita in public transport infrastructure.

The model evaluates an entire investment policy as opposed to a single project analysis, but using the project level cost-benefit micro-economic approach to compare the investment per capita in public transport to the time benefits generated. The model is structured as a combined set of sub models describing the current and long term past investments impact on modal split, speed, and travel time. We develop a city-level aggregate modal split based on city characteristics and public transport investments. This is a different approach than the traditional demand elasticity analysis often used in project level analysis.

Data Sources

We used various data sources and built a combined city-level database. A time series of city-level transport data and modal split data is very limited and often hard to compare. For this research we used these national and city-level data-bases:

UITP Millennium City data base for sustainable mobility—The UITP research database covers 100 cities with more than 200 urban and transport indicators. The research was published in 2001 and includes transport usage by mode, transport supply by mode, energy consumption, financial and cost data, urban network indicators, etc. Although transport data is elaborate, it is limited to a one-year snapshot and does not provide a long-term data series. Table 1 summarizes some main indicators and statistics from the database that were used in this research.

Table 1. Millennium city database main indicators and statistics.

	Average	**Min**	**Max**	**Std**	**units**
Demographic data					
Number of cities	95				
Population	4,662,123	240,066	32,342,698	32,342,698	persons
Urban density	75	6	230	58	persons/ha
Proportion of jobs in CBD	17	3	75	14	%
Supply indicators					
Length of road per 1000 people	2955	148	9595	2487	m/1000 persons
Parking spaces per 1000 CBD jobs	266	3	1883	294	spaces/1000 jobs
Traffic intensity indicators					
Passenger cars per 1000 people	336	8	746	195	units/1000 persons
Passenger cars per kilometre of road	139	25	696	108	units/km
Private passenger vehicle kilometres per kilometre of road	1,508,890	133,178	4,802,213	948,996	v.km/km
Public transport supply and service					

Total length of reserved public transport routes per 1000 people	100	0	727	127	m/1000 persons
Public transport vehicle kilometres of service per capita	78	15	420	64	v.km/person
Buses per million people	707	77	3154	469	units/10^6 persons
Average speed of public transport	23	11	50	9	km/h
Moibility indicators					
Total daily trips per capita	2.7	1.2	4.7	0.9	trips/person
Percentage of motorised private modes over all trips	51	7	97	23	%
Total public transport boardings per capita	248	11	1036	191	boardings/person
Transport financial cost					
Public transport investment per capita	77	1	407	91	USD/person
Road investment per capita	165	2	847	169	USD/person

EPOMM—European Platform on Mobility Management—We use TEMS, the EPOMM modal split tool, a database of 453 cities (mostly European) with detailed modal split data.

The EPOMM member countries complete the best possible data based on as much survey background data as possible. Modal split data of each city is given for a specific year (based on the survey available), mostly from 2006-2013. We use the newer TEMS modal split data where it was available, and complete it with the millennium city database modal split data where it is missing.

See the Appendix for the list of cities with some basic characteristics.

EMPIRICAL ANALYSIS

In the first step we estimate an aggregate modal split model (MSM) that predicts the percent of public transport trips based on the investment in PT and city characteristics. The MSM model is a regression model estimating the share of public transport trips. The model results are shown in Table 2.

The parameter PT Invest_developed is the annual investment in public transport (in US dollars per capita) if the city is in a developed country and zero if the city is in a developing country. The parameter was found positive and significant at the 5% confidence level and shows that investment in public transport contributes to PT usage in the developed world. This parameter was found to be insignificant in the developing world, possibly because of the opposite effects of income increases and investment on modal split.The proportion of jobs in the CBD shows a positive contribution of the strength of the CBD and public transport usage. This parameter is positive but significant only at the 10% confidence level. It was found significant in other variations of the model and given its importance, we decided to keep it in the model.

The past investment in public transport is taken into account by the parameter total PT reserved routes km per passenger, the length of all public transport reserved routes such as metro, suburban rail, light rail, and bus rapid transit (BRT) exiting in each city. This parameter was found positive and highly significant showing the contribution of this variable to modal split.

We found a negative contribution of the motorization level to public transport usage. The parameter passenger cars per 1000 people is negative and highly significant. Finally, the parameter total public transport annual vehicle km of service per capita represents the supply of public transport. This parameter reflects the amount of public transport service provided by the city in terms of line length and frequencies, but has some limitations as it does not necessarily show service coverage and effectiveness. The parameter was found to have a positive contribution to public transport usage and significance at the 5% level.

Table 2. The City-level Modal Split Model (MSM).

SUMMARY OUTPUT				**Model 1**		
				Annual Investmet per capita		
Regression Statistics				new modal split data		
Multiple R	0.74			95 cities		
R Square	0.55					
Adjusted R Square	0.52					
Standard Error	14.78					
Observations	95					
	Coefficients	Standard Error	t Stat	P-value	Lower 95%	Upper 95%
Intercept	42.427	4.689	9.05	2.98784E−14	33.11	51.74
PTInvest_developed	0.037	0.018	2.08	0.040223048	0.00	0.07
Proportion of jobs in CBD	0.181	0.136	1.34	0.185031661	−0.09	0.45
Total PT reserved route km	0.007	0.002	3.87	0.000204436	0.00	0.01
Passenger cars per 1000 people	−0.061	0.009	−7.02	4.2031E−10	−0.08	−0.04
Total public transport vehicle kilometres of service per capita	0.084	0.025	3.33	0.001279612	0.03	0.13

Several modal split models were estimated based on the data available. Another version of model 1 (model 1ln) included the logarithmic of the parameter PT Invest_developed. This version has yielded very similar results. Many other parameters were tested for inclusion in the model, among them city population, population density, GDP, public transport to car speed ratio, PT speed and mass transit speed. They were found to be less significant in the model when combined with the public transport investment parameter, which is the main parameter we are interested in analyzing.

Since modal split is in the range of 0 - 100, we made sure that the results are not outside the range. We also verified that any reasonable range of the input parameters will result in a model prediction within the 0 - 100 range, thus there is no need for some type of censored model.

The next step of the analysis estimates the PT speed model (PTSM) using regression. The model estimates PT average speed based on the accumulated investment (PT inventory value) in public transport, length of PT reserved routes, and length of roads (Table 3). The model is then used to estimate the effect of the additional annual investment in public transport of each city on PT average speed and hence travel time (assuming length of trip remains the same).

The first parameter in the model is the log of the accumulated investment in public transport. The two other parameters which were found to be significant are total public transport reserved route km and total length of road per 1000 people.All the parameters have positive contribution to public transport speed. The model shows that higher investment in public transport increases speed logarithmically up to about 30 km/h, as shown in Figure 2.

Reserved public transport routes allow PT to travel at higher speed and avoid congestion. This parameter was found to be positive as expected. On the other hand, buses use mainly the road network and are affected by the density and congestion. The length of road per 1000 people is a measure of the road supply and capacity and was found to have a positive effect on the public transport speed.

In the next step, we estimated a road network (car) speed model (CSM) that relates the network travel speed to traffic density, using regression (Table 3). This model is needed to measure the change in car trip time caused by the annual investment in public transport and the shift of trips from car to PT estimated by the modal split model (MSM) in phase 2.

The model uses the parameters: road density showing that the reduction in road density will cause travel speed to increase; urban density that was found to have a high negative effect on road network speed, as expected; and km of freeway in the city that was chosen to distinguish cities in America and Australia, typified by high-level freeways and higher network speed, from cities in other world regions.

We estimate the accumulated investment in road and in public transport infrastructure (using today's prices). We used current average cost per km of road and public transport by type of service (urban roads, freeways, metro, suburban rail, light rail, and BRT) to calculate the value of each city transport inventory using today's average construction costs (average cost data is based on Andersson, Gibrand, & Fredriksson [15] and Doll & van Essen [16]). We have also collected data on annual investment in public transport for each city [11] .

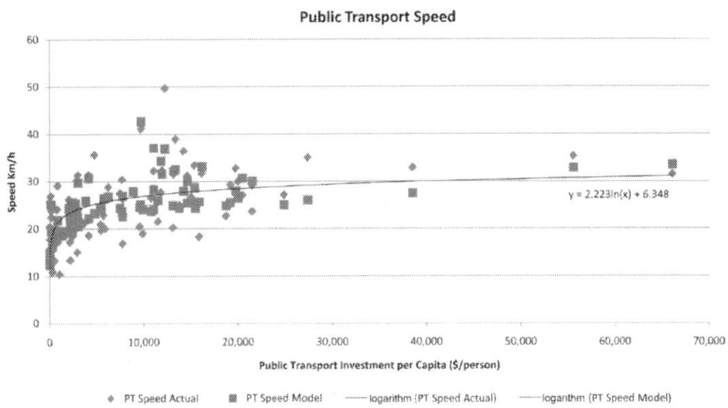

Figure 2. The effect of public transport investment on public transport speed.

Table 3. Public transport speed model and road network (car) speed model.

PT SPEED Model (PTSM)			**Regression Statistics**	
	Coefficients	t Stat	Multiple R	0.75
Intercept	8.8600	3.46	R Square	0.57
Ln(PTInvest)	1.2735	3.89	Adjusted R Square	0.55
Total PT reserved route km	0.0030	4.74	Standard Error	4.96
Length of road per 1000 people	0.0012	4.89	Observations	83
Car Speed Model (CSM)			Regression Statistics	
	Coefficients	t Stat	Multiple R	0.68
Intercept	40.3195	22.11	R Square	0.46
Road density	−0.0046	−1.85	Adjusted R Square	0.44
Urban density	−0.0689	−5.50	Standard Error	7.68
Length of freeway	0.0089	3.01	Observations	89

The last step completes the calculation of the right side of Equation (2):

- Time benefits for existing public transport users based on the change in time estimated by the PT speed model (PTSM).
- Time benefits for private cars users who switch to public transport based on the PTSM model time saving estimates by the rule of half. The number of new PT users is estimated by the modal split model (MSM).
- Time benefits to car users based on road network speed model (CSM) that estimates travel time savings due to the decrease in road density enabled by the shift to public transport.

We then compared the time benefits on the right side of Equation (2) to the actual annual investment per capita data and calculated the time benefit/cost ratio for each city. This analysis examines the actual annual investment in public transport carried by various cities, and aims to show their long-term benefits in terms of time saving. The model takes into account current public transport and road networks in each city and examines the impact of an additional annual investment policy. It should be noted that the model does not show if the accumulated investment (in terms of public transport inventory) is at this point economically justified, but only the current annual investment policy and its benefits.

The results of the models show (Figure 3, Table 4):

- The results show that the worldwide average investment in public transport per capita (inventory value) was almost eight thousand dollars, accounting for 49% of total average investment in transportation. The table also shows big differences in urban transport investments between world regions. While the total investment in transport in Western Europe and America is similar (in the range of 20 thousand dollars per capita), European cities invested on average 65% in public transport while American cities invested only 24% in public transport.

Table 4 shows the differences of main public transport characteristics for cities with high investments in public transport (more than 10 thousand USD per capita) and cities with low investments in public transport (less than 1500 USD per capita). The analysis shows:

- The average speed of the public transport network is almost double in cities with high PT investments. In these cities, an investment of 10,000 USD per capita stimulated an increase in the average speed by 13 km/h (an average of 750 USD per capita per 1 km/h increase in public transport speed). Model results show that investment in PT increases public transport share logarithmically up to about 30 km/h.

- The investment in public transport helped to increase the relative attractiveness of the PT to car use by increasing the PT/car speed ratio from 0.6 to 0.8.
- Public transport share has remained almost unchanged. The reason for this is that cities with low investment in public transport often have a low GDP and low level of motorization, and accordingly high usage of public transport. This result suggests that developed cities that invested in public transport managed to maintain the modal split even with the growth in the level of motorization.
- The investment in PT generates time benefits that cover on average 0.6 - 0.7 of the annual current investment in public transport.

Some cities have B/C ratios higher than 1.0, demonstrating that the time benefits predicted by the model alone cover the investment.

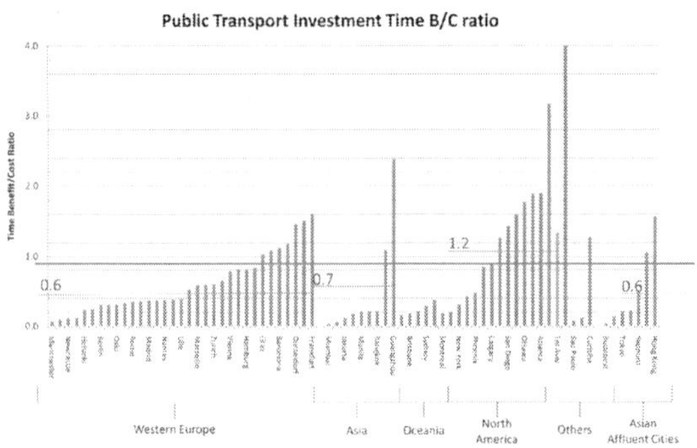

Figure 3. Model results—B/C ratio by world region.

Table 4. Main public transport characteristics for high and low public transport investment cities.

		Low	**High**
	units	<1500 $/person	>10,000 $/person
No. of Cities	no.	31	25
Average PT Investment	$/person	301	10,133

Average PT Speed	Km/h	17	30
PT/Car Speed	ratio	0.6	0.8
Average PT Share	% PT Trips	38	40
Motorization Level	cars/1000 persons	223	422
GDP	$/person	10,179	31,474

- Cities with developed public transport systems, like Western Europe and affluent Asian cities, invest on average over 200 dollars per person per year. The time benefits generated is on average 0.6 of the investment. The results show that cities with already developed transit systems probably only cover the investment. One of the reasons can be the need for high maintenance and upgrade of existing public transport infrastructure that has a limited contribution to additional time saving.
- Middle East cities have reaped high time benefits for current investment policies. This is probably due to the fact that these cities made almost no previous investments in public transport, and the improvement gains at the early stages of development can be significant.

SUMMARY AND CONCLUSIONS

In this research, we focused on urban public transport investments in various cities and examine the relationship between public transport and road network investments, speed, GDP, and modal split. We developed a city-level indicative urban public transport investment model based on micro-economic theory to analyze urban public transport investments and the impact on time saving benefits. The model uses city aggregated data and various sub models to estimate the relationship between public transport investments, speed, modal split, and the ensuing time benefits generated by the investment.

The main results of the model:

- In developed cities, PT investment contributed to the use of public transport. Public transport reserved routes (for PT inventory or past investments), jobs proportion in the CBD, and PT supply were also found to have a positive effect on PT modal split, while motorization levels was found to have a negative impact on PT usage as expected.

- A comparison of cities with high and low investments in public transport reveals some interesting characteristics of urban public transport and road investments. The results showed that cities invested on average 7 - 8 thousand US dollars per capita in public transport, accounting for about 50% of the total transport budget. Cities with more developed public transport system invested about 15 thousand US dollars per capita, and allocated 65% of the budget to public transport. These cities manage to maintain the average public transport speed in the range of 30 km/h (on average a 1.3 km/h improvement in the public transport average speed for every 1000 dollar investment per capita.
- The model estimates time benefits for public transport and road network users generated by the investment in public transport. The time benefits for each city are then compared to the investment in public transport. The results show:
  - Investment in public transport increased PT share.
  - The investment in cities with developed public transport systems generated time benefits that covered on average 0.6 - 0.7 of the investment.
  - Some cities have B/C ratios higher than 1.0, demonstrating that the time benefits predicted by the model cover the investment.

There is limited research regarding the investment in urban public transport and road networks and the right balance between these investments. We believe that this research can help to better understand and plan urban public transport networks. The research can contribute to researchers and policy makers to better direct the level of investment and the expected city level impacts.

The model developed in this research has some limitations. The model does not find an optimal investment policy or a general equilibrium between public and private transport. The model analyzes the impact of past public transport investment on time benefits and determines to what extent these benefits cover the investment. The city-level aggregate model is based on city-level average data. This has some limitations to model predication and analysis. For example, congestion and modal split data between the city center and other areas can be very different. Average modal split does not incorporate all the differences between cities' urban structures and characteristics. Average speed changes are limited to over 30 km/h, so the model is limited in its ability to estimate the additional time benefits in cities with already developed and higher speed public transport systems. Some improvements can be introduced to develop the analysis, such as using peak and off-peak data, and adding environmental, safety, and agglomeration benefits.

Further research is needed to investigate the balance between urban public transport and road investments and develop models that will combine more micro and macro approaches. Research focused on a combined model can further analyze the relationship between city size and density and public transport investment, and the effect of the city residents' preferences on the balance between road and PT investment. City-level data was very limited in this research. There is a need for better worldwide annual data on congestion, speed, modal split and travel costs in cities and city centers to understand more about the impact of transport investments.

ACKNOWLEDGEMENTS

This work presented in this paper was partially supported by the Israel Sceince Foundation (ISF).

APPENDIX—CITY DATABASE

City	Country	Region	Population in Metropolitan area Population	Surface area (ha) Surface	Transport modes						
					Bus	Minibus	Tramway	Light rail	Metro	Heavy rail	Other
Abijan	Ivory Coast	AFRICA	2,790, 000	56,550	X	X					X
Amsterdam	Netherlands	WESTERN EUROPE	831,499	52,020	X		X	X	X	X	
Athens	Greece	WESTERN EUROPE	3,464,866	253,500	X				X	X	
Atlanta	United States of America	NORTH AMERICA	2,897,178	773,530	X	X			X		
Bangkok	Thailand	OTHER ASIAN CITIES	6,685,000	157,725	X					X	X
Barcelona	Spain	WESTERN EUROPE	2,780,342	33,150	X				X	X	
Beijing	PR China	OTHER ASIAN CITIES	8,164,000	456,790	X	X			X		

Berlin	Germany	WESTERN EUROPE	3,471,418	83,351	X		X		X	X	
Berne	Switzerland	WESTERN EUROPE	295,837	31,358	X		X	X		X	
Bogota	Colombia	LATIN AMERICA	5,569,633	173,000	X	X					
Bologna	Italy	WESTERN EUROPE	448,744	20,300	X						
Brasilia	Brazil	LATIN AMERICA	1,821,946	578,916	X						
Brisbane	Australia	OCEANIA	1,488,883	462,068	X					X	X
Brussels	Belgium	WESTERN EUROPE	948,122	16,051	X		X		X	X	
Budapest	Hungary	EASTERN EUROPE	1,906,798	52,516	X		X	X	X	X	X
Buenos Aires	Argentina	LATIN AMERICA	11,355,562	3,880,000	X			X	X	X	
Cairo	Egypt	AFRICA	13,144,000	143,569	X	X	X			X	X

Calgary	Canada	NORTH AMERICA	767,059	72,173	X	X		X			
Cape Town	South Africa	AFRICA	2,900,000	215,520	X	X				X	
Caracas	Venezuela	LATIN AMERICA	4,550,313	37,000							
Casablanca	Morocco	AFRICA	3,094,000	110,000	X						
Chennai	India	OTHER ASIAN CITIES	6,083,371	116,700	X					X	
Chicago	United States of America	NORTH AMERICA	7,523,328	955,710	X	X			X	X	
Copenhagen	Denmark	WESTERN EUROPE	1,739,458	289,187	X					X	
Cracow	Poland	EASTERN EUROPE	744,987	32,880	X		X			X	
Curitiba	Brazil	LATIN AMERICA	2,431,804	1,330,600	X	X					
Dakar	Senegal	AFRICA	1,939,000	55,000	X	X				X	

Delhi	India	OTHER ASIAN CITIES	11,300,000	148,639							
Denver	United States of America	NORTH AMERI-CA	1,984,578	949,690	X			X			
Dussel-dorf	Germany	WEST-ERN EUROPE	571,064	20,300	X		X	X		X	
Frankfurt	Germany	WEST-ERN EUROPE	653,241	24,289	X		X		X	X	
Geneva	Switzer-land	WEST-ERN EUROPE	399,081	24,232	X		X			X	X
Glasgow	United Kingdom	WEST-ERN EUROPE	2,177,400	480,700	X				X	X	X
Graz	Austria	WESTERN EUROPE	240,066	12,609	X		X				
Guangzhou	PR China	OTHER ASIAN CITIES	3,853,800	144,360	X						X
Hamburg	Germany	WESTERN EUROPE	1,707,901	69,393	X				X	X	X
Harare	Zimbabwe	AFRICA	1,432,260	58,682	X	X					
Helsinki	Finland	WESTERN EUROPE	891,056	74,300	X		X		X	X	X

Ho Chi Minh City	Viet nam	OTHER ASIAN CITIES	4,811,170	209,370	X	X					
Hong Kong	PR China [HKSAR]	ASIAN AFFLUENT CITIES	6,311,000	109,591	X	X	X	X	X	X	X
Houston	United States of America	NORTH AMERICA	3,918,061	1,737,550	X						
Istanbul	Turkey	EASTERN EUROPE	9,076,865	551,200	X	X	X	X		X	X
Jakarta	Indonesia	OTHER ASIAN CITIES	9,161,000	66,168	X	X				X	
Johannes-burg	South Africa	AFRICA	2,448,436	138,400	X	X				X	
Kuala Lum-pur	Malaysia	OTHER ASIAN CITIES	3,773,900	285,337	X				X	X	
Lille	France	WESTERN EUROPE	1,153,000	87,900	X			X	X	X	
Lisbon	Portugal	WESTERN EUROPE	2,556,180	312,800							
London	United King-dom	WESTERN EUROPE	7,007,100	157,900	X			X	X	X	
Los Angeles	United States of America	NORTH AMERICA	9,077,853	1,051,530	X			X	X	X	
Lyon	France	WESTERN EUROPE	1,152,259	48,675	X	X			X	X	
Madrid	Spain	WESTERN EUROPE	5,181,659	802,790	X				X	X	
Manchester	United King-dom	WESTERN EUROPE	2,578,300	127,200	X			X		X	
Manila	Philippines	OTHER ASIAN CITIES	9,447,156	63,780	X	X		X		X	

Marseille	France	WESTERN EUROPE	798,430	23,850	X		X		X		
Melbourne	Australia	OCEANIA	3,138,147	769,086	X		X			X	
Mexico City	Mexico	LATIN AMERICA	15,748,038	497,400	X	X		X	X		
Milan	Italy	WESTERN EUROPE	2,460,000	62,900	X		X		X	X	
Montreal	Canada	NORTH AMERICA	3,224,130	341,164	X				X	X	
Moscow	Russian Fed	EASTERN EUROPE	8,700,000	109,100	X		X		X	X	
Mumbai	India	OTHER ASIAN CITIES	17,072,000	423,600	X					X	
Munich	Germany	WESTERN EUROPE	1,324,208	30,656	X		X		X	X	
Nantes	France	WESTERN EUROPE	534,000	48,700	X			X			
New York	United States of America	NORTH AMERICA	19,227,361	2,276,340	X			X	X	X	X
Newcastle	United King-dom	WESTERN EUROPE	1,131,000	54,014	X				X	X	X
Osaka	Japan	ASIAN AFFLUENT CITIES	16,828,737	1,489,100	X			X	X	X	
Oslo	Norway	WESTERN EUROPE	917,852	501,300	X		X		X	X	X
Ottawa	Canada	NORTH AMERICA	972,456	484,000	X						
Paris	France	WESTERN EUROPE	11,004,254	1,201,200	X			X	X	X	
Perth	Australia	OCEANIA	1,244,320	538,686	X					X	X
Phoenix	United States of America	NORTH AMERICA	2,526,113	2,383,850	X	X					

Prague	Czech Republic	EASTERN EUROPE	1,212,655	48,442	X		X		X	X	
Rio de Janeiro	Brazil	LATIN AMERICA	10,192,097	5,737,800	X		X	X	X	X	X
Riyadh	Saudi Arabia	MIDDLE EAST	3,116,000	146,800	X	X					
Rome	Italy	WESTERN EUROPE	2,654,187	128,530	X		X	X	X	X	
Ruhr	Germany	WESTERN EUROPE	7,356,500	488,759	X		X	X		X	
Salvador	Brazil	LATIN AMERICA	2,663,481	247,780	X					X	
San Diego	United States of America	NORTH AMERICA	2,626,714	1,088,960	X			X		X	
San Francisco	United States of America	NORTH AMERICA	3,837,896	640,600	X		X	X	X	X	X
Santiago	Chile	LATIN AMERICA	5,090,914	226,700	X				X	X	
Sao Paulo	Brazil	LATIN AMERICA	16,562,227	805,100	X				X	X	
Sapporo	Japan	ASIAN AFFLUENT CITIES	1,757,025	112,212	X		X		X	X	
Seoul	South Korea	OTHER ASIAN CITIES	20,576,272	1,174,792	X				X	X	
Shanghai	PR China	OTHER ASIAN CITIES	9,570,000	205,701	X						
Singapore	Singapore Republic	ASIAN AFFLUENT CITIES	2,986,500	64,750	X	X			X		
Stockholm	Sweden	WESTERN EUROPE	1,725,756	649,000	X			X	X	X	X
Stuttgart	Germany	WESTERN EUROPE	585,604	20,462	X		X	X		X	

Sydney	Australia	OCEANIA	3,741,290	1,214,385	X					X	X
Taipei	Taiwan	OTHER ASIAN CITIES	5,960,673	232,437	X					X	
Tehran	Iran	MIDDLE EAST	6,800,000	88,170	X	X					
Tel Aviv	Israel	MIDDLE EAST	2,458,155	151,900	X					X	
Tokyo	Japan	ASIAN AFFLUENT CITIES	32,342,698	1,355,276	X			X	X	X	
Toronto	Canada	NORTH AMERICA	4,628,883	717,491	X		X		X	X	
Tunis	Tunisia	AFRICA	1,874,600	260,000	X			X		X	
Turin	Italy	WESTERN EUROPE	1,451,000	61,200	X		X			X	X
Vancouver	Canada	NORTH AMERICA	1,898,687	282,066	X				X	X	X
Vienna	Austria	WESTERN EUROPE	1,592,596	39,528	X		X	X	X	X	
Warsaw	Poland	EASTERN EUROPE	1,628,500	49,500	X		X		X	X	
Washington	United States of America	NORTH AMERICA	3,739,330	610,580	X				X	X	
Wellington	New Zealand	OCEANIA	366,411	164,100	X					X	X
Zurich	Switzerland	WESTERN EUROPE	785,655	62,582	X		X	X		X	X

Source: Vivier, J., Kenworthy, J., & Laube, F. (2001). Millenium Cities Database for Sustainable Mobility. UITP.

REFERENCES

1. Newman, P. and Kenworthy, J. (1999) Sustainability and Cities: Overcoming Automobile Dependence. Island Press, Washington.
2. Berechman, J. (2009) The Evaluation of Transportation Investment Projects. Routledge, Abingdon-on-Thames.
3. Hayashi, Y. and Morisugi, H. (2000) International Comparison of Background Concept and Methodology of Transportation Project Appraisal. Transport Policy, 7, 73-88.
4. Mackie, P. and Worsley, T. (2013) International Comparisons of Transport Appraisal Practice. Institute for Transport Studies, University of Leeds, Leeds.
5. Vickerman, R. (2000) Evaluation Methodologies for Transport Projects in the United Kingdom. Transport Policy, 7, 7-16.
6. Shiftan, Y., Sharaby, N. and Charles, S. (2008) Transport Project Appraisal in Israel. Pricing Economics and Finance. Transportation Research Board No. 2079, 136-145. https://doi.org/10.3141/2079-17
7. Venable, A.J. (2007) Evaluating Urban Transport Improvements: Cost Benefit Analysis in the Presence of Agglomeration and Income Taxation. Journal of Transportation Economics and Policy, 41, 173-188.
8. Melo, P. and Graham, D.J. (2009) A Meta-Analysis of Urban Agglomeration Economies. Regional Science and Urban Economics, 39, 332-342.
9. Mackie, P., Graham, D. and Laird, J. (2011) The Direct and Wider Impacts of Transport Projects: A Review. In: De Palma, A., Lindsey, R., Quinet, E. and Vickerman, R., Eds., A Handbook of Transport Economics, Edward Elgar Publishing, Cheltenham, 501-526. https://doi.org/10.4337/9780857930873.00030
10. Banister, D. and Berechman, J. (2000) Transport Investment and Economic Development. Routledge, London.
11. Vivier, J., Kenworthy, J. and Laube, F. (2001) Millenium Cities Database for Sustainable Mobility. UITP. [Citation Time(s):3]
12. Mogridge, M. (1997) The Self Defecting Nature of Urban of Urban Road Capacity Policy—A Review of Theories, Disputes and Available Evidence. Transport Policy, 4, 5-23.
13. Basso, L.J. and Jara Diaz, S.R. (2012) Integrating Congestion Pricing,

Transit Subsidies and Mode Choice. Transportation Research Part A Policy and Practice, 46, 890-900.

14. Beaudoin, J., Farzin, H. and Lin, C. (2014) Evaluating Public Transit Investment in Congested Cities.
15. Andersson, P.G., Gibrand, M. and Fredriksson, L. (2009) Proceed, Principles of Successful High Quality Public Transport. Directorate General for Energy and Transport.
16. Doll, C. and van Essen, H. (2008) Road Infrastructure Cost and Revenue in Europe. CE Delft.

SECTION IV
CASE STUDIES FROM DIFFERENT COUNTRIES

CHAPTER 14

Transport Policy—The Perspective of Swedish Stakeholders

Hilda Karlsson, Rickard Bergqvist, Lars Brigelius

Industrial and Financial Management & Logistics, Department of Business Administration, School of Business, Economics and Law at University of Gothenburg, Göteborg, Sweden

ABSTRACT

Differences between countries' internalization degrees and between transport modes both tend to risk distorting competition and creating socioeconomic inefficiency. This review examines regulations and charges for freight transports for different transport modes in Sweden and compares the situation with other key European countries. The aim was to identify best

Citation: A. Karlsson, H. , Bergqvist, R. and Brigelius, L. (2014), "Transport Policy—The Perspective of Swedish Stakeholders". *Journal of Transportation Technologies*, **4**, 175-186. doi: 10.4236/jtts.2014.43018.

practices and benchmarking for Swedish conditions. The project involved both a literature study and interviews with different stakeholders. It was found that today's fee structure for goods transports often deviated from the views of different stakeholders and market segments. The problem is that it is difficult to estimate and plan well in advance due to a lack of relevant data. Economic transport data are often extracted from heavily aggregated data where resource consumption and production costs for transport are unclear, preventing reliable estimates and obstructing the calculation of marginal costs and internalization of externalities through charges imposed on the respective transport modes and transport units. Furthermore, there is little research on how railroad and shipping are affected by changes and introduction of truck fees. The latter implies that truck fees must be seen from a European perspective to assess their effect on the overall transport system. Other aspects that are important to highlight are the extent and point in time for railroad deregulation. Further, it became clear from the stakeholder analysis that all fees and policies need to be both comprehensible and transparent through good communication of rules and by explaining the fee structures to all those concerned.

Keywords: Transport Policy, Fees, Regulation, Stakeholder Analysis, Sweden

INTRODUCTION

As illustrated in past research and studies by Trafikverket [1] and Nash et al. [2] , differences between countries' internalization degrees and differences between transport modes both tend to risk distorting competition and creating socioeconomic inefficiency. Possible improvement proposals given by the interviewed respondents have the possibility to influence long-term predictability and transparency among regulators, capacity evolution and capacity allocation. Many of the respondents emphasized current shortcomings related to the condition and maintenance of transport infrastructure. Furthermore, the current fee structure is often viewed as complex and difficult to grasp by many industry players. The latter interviewed respondents also pointed out that the differences in freight fee levels between different transport modes could distort competition. Proposals that have emerged in the study are successive capacity allocation for railway capacity, public-private partnership (PPP) for infrastructure investments and a more quality-based and differentiated fee system. The bottom line is that more knowledge about the transport policy, fee systems, and external costs of transport is required for a better differentiated fee system and a higher degree of internalization of external costs.

Fees and regulations for the carriage of goods are current and controversial subjects that require more and more knowledge as the demand for transport increases and transport policy frameworks for freight transport get more restrictive and complex. The aim of this paper is to identify impairments and best practices related to freight transport policy in Sweden as compared to other European policies by means of a literature review and stakeholder analysis. Special attention is given to road and rail transport policy and regulatory issues.

METHODOLOGY

The context of this study has been developed by means of a literature review of previous research and stakeholder interviews. Special attention has been given to the European regulatory framework related to rail transport and to that of countries with especially predominating rail infrastructure such as Germany, Switzerland, Austria, The Netherlands and France. These countries have been chosen based on many reasons, e.g., they have truck fees, support the development of rail services by subsidies, and/or have liberalised rail services. Switzerland, Austria and Germany have been especially interesting to study since these countries imposed truck fees early and their effects have been thoroughly analysed.

The stakeholder analysis addresses different stakeholder perspectives and opinions about the fees, rules and suggestions for improvement. The study begins with a description of the transport market development in Sweden. Then Sweden is compared with other EU countries' transport markets. The study also looks at the determinants of the transport system stakeholders, while the reference frame provides an inventory and review of policy, fees and structures.

The study applies semi-structured interviews, which means that all respondents are asked the same questions, with open response options [3] . Thus the respondents are treated equally and they have an opportunity to comment on the same questions. The selection method for the respondents is based on the research questions and the purpose is to obtain a broad and thorough description of the problem area. For this purpose, different categories of respondents and experts in Sweden were selected, comprising authorities, industrial organizations, and infrastructure users. Since the purpose of the interviews is not to get a statistical generalization but rather detailed descriptions, the size of the respondent population was kept low in the study and no more respondents were interviewed after the point at which new major observations ceased to be obtained. Before the interviews were conducted, an information letter was sent in advance, containing a personal presentation and information about the purpose of the interviews, the confidentiality of the results, the voluntary nature of their participation and the possibility of subsequently withdrawing their participation, contact information in case of questions or concerns, how the results will be published, how the respondent can take part init and about possible additional questions that may be asked after the interviews.

The interview guide with the selected questions was tested in a sample interview, which was recorded and analysed. Ethical aspects considered were that answers from the interviews would only be published after the respondent's approval and were therefore not liable to harm the respondents or risk their identification. The survey was presented openly and honestly through an information letter sent in advance of the interviews. It was clear that any participation was voluntary and that an acknowledgment from the respondent was required prior to his participation in the study. The different categories of stakeholders were politicians, policy makers, transport authorities, transport regulatory agencies, transport industry associations, environmentally oriented NGOs and transport service providers. A total of 18 interviews were conducted with 16 different stakeholders/organisations.

THEORETICAL FRAMEWORK

The demand for freight transport is determined largely by economic development. A weak decoupling between these aspects has been reached, as can be seen in the GDP growth outpacing the freight transport volume over time [4] . However, increasing the efficiency of vehicles and vessels is not sufficient—the efficiency of the transport system as a whole also must be improved. It is in this context that transport policy plays an important role as both an enabler and a facilitator, connecting individual transport movement demands with overall society goals and objectives. The theoretical framework is structured into three main parts comparing railway track fees, truck fees and the degree of internalisation in Europe.

A comparison of track fees between Sweden and Europe The Swedish track access charges are low from an international perspective. The track charges' share of rail services costs in 2009 in Europe were 30% for freight and the corresponding figure for Sweden was 6% [5] . [6] has illustrated the widespread fee levels in Europe. Former Eastern European countries charge high fees in comparison to Sweden. In 2006, track charges were abolished for intermodal traffic in Norway, reducing the average cost. Track access charges have varied in France but are expected to increase in the future. In this context, one should also note that infrastructure charges in Sweden have gradually increased up to the year 2011 and have stayed at the same level since that time.

Differences in infrastructure cause different costs for wear. This can explain the spread of fee levels across countries. According to [7] , the marginal cost variation could depend on the infrastructure quality and the corresponding traffic volume. Rail track fees are also regulated by the politicians' commitment to and prioritization of rail transport in conjunction with the allocation of resources and investment in infrastructure. Countries with marginal cost pricing, however, tend to have track access charges that are too far below the real marginal cost, since the cost of wear and reinvestment has not been included in this pricing [6] . Sweden has low fees, but at the same time it has high transport costs and a limited tax levy for truck transports. Environmental and climate objectives require that the rail freight transport takes over transport volume from road transport and shipping. With a constant or rising transport demand, this leads to a constant or increasing rail freight shipping volume. This explains the caution in increasing the track access charges for freight during the recent years. A low infrastructure charge is a form of government subsidy for rail transportation. In other countries, similar government subsidies exist in different forms [5] .

For Sweden, the total track charges, calculated using the 2009 price levels, are expected to rise from 509 million SEK in 2009 to 2 billion SEK in 2020 [1] . However, the estimated revenues may differ from the prognosis, as higher fees can affect supply and demand. The passage charge in large cities can create incentives to evade the charging system by choosing a place of departure and/or arrival outside the chargeable period. The actual earnings may also be lower than expected as a result of differentiated emission charges for diesel traffic. If the internalization rate for rail freight increases, this is expected to lead to a shift of freight from railway to road transport and the non-internalized external effects will be greater for road transports as compared to what they previously were for railway transports. A reduction of the external effects of railway transports can cause a significant increase in the corresponding effects for road transports, which makes it appropriate to compare the internalization rate by rail and road and to frame the absolute effects in relation to taxes.

A comparison of road user charges for trucks in Europe One of the major European events from a toll perspective took place on January 1, 2005, when truck fees were introduced for selected congested highways and on highways in Germany. The fee in Germany is regulated by the amount of emissions and the number of axles [8] .

During the years 1993-2003, Austria had a so-called Ökopunkte system that would restrict transit with a number of points each year which were allocated by both Austria and the EU. The aim was to gradually reduce emissions of NOx by 60% by 2003. This would have been achieved by allocating a limited number of points, based on their NOx emissions, to vehicles with a gross weight of over 7.5 tons. The Ökopunkte system was not compatible with EU laws and Austria's entry into the EU contributed to the discontinuation of the system. In 2004, a truck toll for vehicles with a gross weight of over 3.5 tons was introduced. The special extra fee on some selected highways in the Alps is significantly higher than the toll. A time differentiation of the charge means a 50% higher fee between 22.00 - 05.00. This has been introduced on the Brenner motorway, where the specific charges are highest [1] [8] [9] .

Until 2001, there was a vignette system in Switzerland, similar to Eurovignette in Europe although more expensive, for vehicles with a gross weight of more than 3.5 tons. On January 1, 2001 the system was replaced by LSVA, Leistungsabhängige Schwerverkehrsabgabe, for the entire road network, including trucks with a gross weight exceeding 3.5 tons. Austria and Germany, however, impose the truck charges only on certain parts of the network and their calculations differ from the ones used in Switzerland. In Switzerland the charges are counted in ton-kilometres by emission class per vehicle or by the vehicle's maximum permissible gross mass. The charge was just over 1.0 cents/ton-km in fee category II and the truck's allowable gross weight was increased from 28 to 34 tons in 2001. The permissible gross weight was increased to 49 tons and the charge was subsequently increased to 1.6 cents/ton-km on January 1, 2005. Subsequently, an increase in the fee to around 1.7 cents/ton-km was carried out on January 1, 2008. The prohibition against driving a truck at night from 22.00 to 05.00 in the morning remained.

In 1994 the Netherlands, Belgium, Germany, Denmark and Luxembourg formed the Eurovignette, which introduced a common fee for trucks on country highways. In 1997, Sweden joined the Eurovignette by an agreement in accordance with the European Council guideline 1993/89/EWG from

October 25, 1993 and renewed in 1999 by EC Directive 1999/62/EC [10] . The charge is time-dependent, is collected at the use of the road network and differs depending on the number of axles and environmental performance of the vehicle. Since that time, Germany has introduced distance-based charges and has therefore left Eurovignette, as EU law does not allow both time-and distance-dependent charges in parallel. The Eurovignette charge is relatively low compared with recent years' distance-related toll in Europe. The Eurovignette fee is 1250 €/year and corresponds to a cost of 1.25 cents/ mile based on a Euro II/III truck with five axles running 100,000 km/year on the highway. It is comparable to the distance-dependent charge in Germany, which is 10 times higher. Other European countries that have introduced charges on certain parts of the network are Italy, Spain, and France. Truck fees have been decided or have already been introduced in several countries in Europe.

A comparison of the degree of internalisation in Europe The internalization rate differs between modes of transport and it varies between different countries. This is a consequence of, for example, the proportion of highway in relation to the whole road network, and how large the infrastructure wear is. Both of the latter factors are country-specific. Shipping has a significantly lower internalization rate than land-based transports. Higher internalized fees and taxes would be needed for shipping in order to internalize the cost of air pollution to a greater extent than currently. The external road transport cost varies between countries. Countries with road tolls such as Germany, Italy and Austria have a higher degree of internalization. Also, tolls for bridges, for example in Sweden, Norway, and Denmark, affect the internalization rate. High tolls and low external costs may explain the high internalization rates in Austria. The internalization rate for railway transports differs depending on the country. Sweden is characterized by a relatively low internalization rate [11] .

Concerning external costs, more knowledge exists about road transports than about other modes of transport. This may also explain the difference in fees and taxes. For shipping, for example, restrictions on allowable sulphur levels in fuels are used as a substitute for taxation. Other examples include the railway route Oslo-Rotterdam, which has a higher internalization rate than the route Narvik-Napoli because of the fee system for the tracks [11] . Wear is one of the major factors characterizing the external costs of rail. To reach a full internalization for railway tracks, a review of the fee system that takes into account the external costs is needed. Today's methods for calculating the cost of wear are based solely on maintenance costs. However, the operation and development costs also should be components in the calculation of costs of wear. The internalization for road transports includes some operational costs and development costs, which makes a comparison with the corresponding figures for rail difficult. The largest part of the total external costs for road and rail is the cost of wear. For full internalization, more knowledge about the wear of the infrastructure is needed, as well as a review of taxes within and between countries. The share of highways in relation to all roads, road tolls, and bridge tolls affects the external costs and the internalization rate. Other factors that may contribute to increased external costs are noise and congestion. The use of diesel locomotives as substitutes for electric locomotives affects non-internalized external costs that can also be used in calculating the degree of internalization [11] .

Making international comparisons is complex, since the design and structure of taxes and fees may vary by country. An example of this is the Swedish railway charges, which are disaggregated unlike those in the compared countries, which have chosen to integrate all external costs in a track access fee. The knowledge of external costs also differs between countries and transport modes, which complicates a deeper understanding of how the fee system is designed. The calculation methods for external costs vary between different modes of transport and the uncertainties with regard to the calculation of external costs pose a risk for misleading results [11] .

For an internalization rate lower than 100 percent, the additional cost to be carried by the society is the remaining non-internalized cost of externalities. A comparison between modes shows the charge level needed for full internalization. Today's measurement problems mean that some externalities are not included in the calculations, which may result in an overestimation of the internalization rate and, to some extent, an underestimation of the economic costs. It is an incentive to work for better documentation and monitoring of the external effects, thereby overcoming the problems. This would lead to more reliable results and a more accurate and appropriate fee level. Road traffic generally has a higher degree of internalization.

Socio-economic efficiency is achieved when the user's total cost is equal to the socio-economic cost of infrastructure use. The sum of the marginal costs of infrastructure use and its external effects should thus amount to the sum of the rates and charges for the use of infrastructure and other external effects. Full cost recovery through marginal cost pricing is not achievable if the total average cost per unit for the use of the infrastructure, including fixed costs, exceeds the marginal cost for using the infrastructure.

Ramsey's rule can be used as an alternative if full cost recovery is not feasible. Ramsey's rule sets prices or fees for the use of the infrastructure that exceed the marginal costs, provided that the distorting effect will be the same for all modes of transport. A decrease in volume due to an increase in prices or fees can have a distorting effect in relation to the first-best, that is, marginal cost pricing. To circumvent this problem, the increase should affect the usage of infrastructure by users with low price sensitivity. Split tariffs with a fixed charge and a marginal cost-based variable fee is an alternative form of second-best pricing. The port fees based on price discrimination are an example of one implementation of the Ramsey rule [12] . [13] shows that marginal cost pricing can be applied with full cost recovery, for example for centrally located ports with congestion problems. External costs can be reduced by alternative fuels and lower speeds for trucks. Reduced road wear, fewer accidents, and less air pollution can be achieved through more efficient and safer vehicles and infrastructure, combined with better community planning with an emphasis on the first step of the four-stage principle. This can be expected to lead to a reduction of the total and the remaining marginal cost of externalities.

Other suggested measures include international collaborations such as the EU emission trading scheme (ETS), changes in the European Transport Directive (ETD) and the sulphur directive regulates the level of sulphur emissions for shipping and affecting international competition [14] . The cost increase may lead to a shift from maritime transport to land transport and possibly have a side effect on carbon emissions. Whether the level of investment, operation and maintenance is sufficient partly depends on the varying cost levels in different countries. Limited resources require a more efficient use and proper priorities. A proposed alternative solution is to examine whether the resource utilization meets society's demands on the transport infrastructure, and to what degree this meets the transport policy objectives. In this case, accessibility problems such as lack of capacity should be taken into account by using different instruments. An example would be high-traffic routes where higher or differentiated fees may apply. The purpose of such fees is to reduce or shift the demand to those parts of the transport system that have a lower load. It could also lead to a skewed distribution as the result of lower accessibility, because a higher utilization rate creates even more strain from an environmental perspective.

Proposed measures that release and increase the capacity for a specific mode of transport can be developed through a point of view covering all modes of transport and their users, and through a more needs-based work.

For optimal freight transports, intermodal solutions tailored to the suitability of the transported goods for the respective modes are needed. Different goals require different instruments and whether the instruments are effective or not with regard to the internalization rate may be difficult to assess. Economic instruments have a gradual effect, while users of the transport system are able to adapt to them in a way that is cost effective for them. A socio-economically efficient and sustainable transport system requires better planning by updating and adapting to new conditions. In the current situation, it is difficult to assess the possibility of achieving the transport policy objectives in the medium term. The transport system is facing major challenges with regard to the varying demographic situations and the increasing population concentration, which will require new innovative logistics solutions. Both challenges and opportunities will be created through greater international integration of Sweden's transport network, several common instruments for EU member states and a common transport policy.

Special fees may be charged provided that they meet a socio-economically efficient use of infrastructure and that they are competitively neutral. The Swedish Transport Administration is working to differentiate the specific charges based on the current capacity utilization. An example is the differentiation of the track access fees with the fee reflecting the current capacity utilization. Similar charges are levied by tolls in major cities, e.g., Stockholm, Gothenburg and Malmö, on non-holiday weekdays and rush hours. The purpose of differentiated fees has been to relieve the tracks with limited capacity and to get a usage spread. This is expected to result in a more efficient use of railway infrastructure and to increase the overall capacity of the transportation system. In the future, based on the current capacity situation, there is a need for more differentiated services and prices partly as an instrument for traffic management, partly in order to increase customer satisfaction and partly for a better adaptation to different market segments. [1]

In the past, a track access charge was levied to cover costs and to free up capacity in the railway network in Sweden. The fee remained constant until 2010. From 2011, a so-called high-level fee was charged on some paths. A crossing charge was introduced in 2011 in three cities, and the performance of the network determined the amount of the charge in combination with demand [15] . Along with globalization and an increasingly international commerce since the early 2000s, the demand for freight transport increased. West, South and North Main Line has had the biggest increases in terms of the number of trains. Today, there is a lack of capacity which causes congestion [1] .

STAKEHOLDERS' PERSPECTIVE

The stakeholder analysis is structured according to the observed main points of view emphasized by stakeholders.

International cooperation and harmonization

Fees and rules should be designed in an international context, while major international collaboration should be used to design charges in a correct way.

The respondents were generally positive about the work on charges and rules and believe that these are needed as instruments in the transport sector. The respondents are in favour of instruments that are used to reduce dependence on energy and fossil fuels and reduce the levels of greenhouse gases and other emissions. However, it appeared from the respondents' answers that Sweden should take a closer look at international developments in this area. Respondents also felt that the instruments should be harmonized across the EU, thereby providing a consistent taxation and governance. Similarly, policy instruments should be harmonized within Sweden by equivalent rules in all municipalities and regions.

Respondents also pointed out the possibility of implementing "soft system" measures. An example of positive control measures are subsidies for the introduction of new technology solutions. Other examples include access to public transport lanes for freight traffic, increased loading zones and similar measures for newer vehicles with higher fill rates. One suggestion is to use various designated "corridors" for this type of positive, "soft system" regulation.

It was also found that the existing work regarding fees and regulations for the carriage of goods is bureaucratic and inefficient. The charging system is described as ineffective and the reason is considered to be a hierarchical approach, divided into several parties: the EU, the Swedish Transport Administration, the Swedish Transport Agency and the Swedish Maritime Administration.

Knowledge and awareness about the fees and rules

Some respondents point out that politicians and authorities have limited knowledge when it comes to the formulation of rules and fees. Respondents emphasize the importance of information and the spreading of information in order to improve users' knowledge of fees and rules for freight. Professional users generally have a good knowledge of fees and rules, partly because of the knowledge that they already possess and partly because they receive information from industry associations and government agencies. Respondents consider that the need for clearer communication is important. According to the respondents, the information on fees and regulations could be improved by a common information portal for agencies in the transport sector and by avoiding so-called special rules.

The transport operators believe they have a good knowledge of the charges and rules but insist that it is an administrative burden. It appeared, however, that it is particularly difficult to reach foreign haulers, although domestically it should not be a problem. Generally speaking, the haulers have very good knowledge of the charges and rules, according to the respondents. However, a better dialogue between operators and regulators is needed because the decisions are sometimes deficient, according to information that emerged from the interviews.

Design of regulations

The overall responsibilities for the work on the design of the regulatory framework should, according to several respondents, lie with the Swedish Transport Agency. The authorities can help by giving directives on what levels of fees and taxes are applicable and how and by whom they should be charged. They also stressed the importance of transparent and fair rules that also take into account the users' viewpoints. Respondents requested increased cooperation between government agencies, industry organizations and other relevant stakeholders. In order to create a better synergy, the wish was expressed that all players in the industry should take responsibility and be involved in the process. The importance of creating a greater customer orientation with more client meetings and joint investigations was highlighted by a respondent who also felt that the government should take a greater responsibility in these matters. One idea put forward was that it would require an industry forum to start a dialogue, even if it is simultaneously the authorities' responsibility to listen to stakeholders.

According to one respondent, the current rules are outdated and they are rarely replaced, but instead are constantly supplemented with new ones. According to several respondents, there is a need for an actor who takes overall responsibility for contributions from a multimodal perspective. In other words, there is a need for more clearly defined roles. The view also emerged that the design of the Swedish regulatory framework must take place in an international context, regardless of transport mode. In addition, there should be transparency and uniformity in the design of regulatory frameworks and the different modes of transport.

An experience that emerged from the EU level was that the users are highly involved in the decision processes and are consulted constantly by both the European Commission and industry associations. Often industry organizations present their proposals as alternatives to the Commission. The coordinated responsibility between different actors can be improved through regular exchange of information but it is important to also take into account behavioural patterns and effects. We also need a clearer link between infrastructure, technologies and business models.

More information and communication is required to illustrate the problems and objectives that the government wants to obtain or achieve with the regulations. The problem of too-fuzzy interference and distribution of responsibilities should be clarified by information. If possible, a more common, national and international regulatory framework should also be developed, according to several of the respondents. This becomes more important as the transport market is becoming more international and more foreign haulers are operating in the country.

One or more authorities should take full responsibility for simplifying the rules. Respondents mention the Swedish Transport Agency, and/or the industry's rule committee or an entirely new agency as examples of agencies that could take on this responsibility. Although it is the responsibility of the entire state administration to push for regulatory reform, it should possibly be driven, for example, by a special government commission, according to the respondents.

The bottom line, according to respondents, is that the EU and other international bodies as well as government and parliament together with the Swedish Transport Administration, the Swedish Transport Authority and the Industry and Commerce rule committee could contribute most to simplify the regulations. The players should have different roles depending on the tasks. The Swedish Transport Administration participates and is actively working with standardization and functions of the systems which can help to simplify regulations. According to the respondents, the European Commission's Transport Directorate, DG Move, plays a central role with regard to the cross-border and international character of the freight sector. Furthermore, the Swedish Transport Administration has the responsibility for long-term planning of the entire Swedish transport system and thus an intermodal competency.

Improvement opportunities related to charges and rules

According to the respondents, clarity and transparency in the regulatory framework is required to the extent that fees are applied, which currently is not always the case. Furthermore, the respondents underline the importance of users being able to easily see how the fees affect the final cost, which requires a clear pricing schedule.

It also emerged that there is a need for more distance-based charges and a better handling of wear and external costs. The respondents suggested that the costs should be better allocated depending on transport modes and external costs.

According to the operators, the authorities should take overall responsibility and, instead of distance-based taxes, impose fees depending on the vehicle's environmental classification. Operators are generally negative towards the distance driven-based tax.

Regulatory compliance

Several respondents emphasize the challenges and problems involved with the regulatory compliance sometimes being low. Exceeding the maximum load weight of trucks was named as an example. According to the Swedish Transport Agency, rules cannot be written in another way to increase compliance. But compliance can be increased by focusing more on the purpose of the rule. This could motivate users and create acceptance and understanding for the application of the rules.

The respondents believe that it is important to analyze and show how the rules affect transportation costs. It is also important to show that the rules are relevant and that the users can influence the framing of the charges. Several respondents pointed out that the formulation of charges and rules can be improved by highlighting and promoting good practices.

Infrastructure and capacity allocation

The industry should have a greater influence on infrastructure posture and capacity allocation in the most important parts of the transport network, according to several respondents. One suggestion is to create specific organizations around the EU's TEN-T core network and around the so-called Green Corridors. Smarter pricing could optimize the use of the existing infrastructure. Likewise, more alternative financing, such as private initiatives and co-financing of municipalities in the development of new infrastructure should be considered. This creates a need for greater cooperation between the public and private sectors and a shorter execution time.

According to several respondents, the slow freight trains should be separated from the fast passenger trains to increase the capacity of the rail system. One respondent pointed out that Sweden needs to invest more in rail in general, but also in high-speed rail passenger service in particular to free up space for cargo.

The problem, according to the respondents, is that the supporting infrastructure is not properly designed, while at the same time the instruments are ineffective. One example that respondents raised was that wheel lathes exist only in a few places in the railway system. The result is that vehicles are moved long distances to get this service performed. The alternative would be to invest or encourage investment in new facilities with wheel lathing. Another similar example is the location of depots. These examples illustrate the importance of a holistic approach when considering an infrastructure system.

Selection and use of infrastructure

Sweden has made significant increases in funding for the development, operation and maintenance of the transportation system. Despite these increases, several respondents highlight that there are major deficiencies in the transportation system, primarily in the railroad system. This affects the choice of transport mode and in the long term it poses the risk of certain modes being considered as less reliable, making it a challenge to use this transport for strategically important flows.

There seems to be a consensus that the railway should be given higher priority on the political agenda. Based on this opinion the respondents reasoned that there is a need for a long-term perspective when it comes to the rehabilitation and development of the railway. A question that emerged was whether the rules and infrastructure have to be harmonized and promote efficient transport between countries. If the railway had been used more economically and more efficiently in a way that benefits the environment, it would have benefited the Swedish transport sector. The respondents believe that it is vital to create incentives by demonstrating the effects of actions, thus creating a desire to contribute, since "everyone must join the track".

Most respondents believe that overall the vehicles registered in Eastern European countries and used in international industry should be replaced. Further, they say that even foreign haulers should pay the congestion charges to prevent a distortion of competition. They argue that there has never been a common regulatory framework for international freight transport. Therefore, they believe that an international framework for emissions allowances should be made for the carriage of goods as soon as is practicable.

Several respondents believe that it is possible to influence the choice of modes of transport by more monitoring and evaluation with the aim to raise awareness of various modes of transport in general and of intermodal transport in particular. In contrast, they argue that it will take more rules at the EU level to control conveyors and move them towards more sustainable transports. More implicitly, it is about being able to affect the rules for cabotage, kilometre and track fees and information for transport buyers. Cabotage can be improved by clearer rules, but some improvement is already visible, according to the respondents. Furthermore, they believe that rules without the possibility of adequate penalties for violations are a problem.

Differentiation/internalisation of fees

An interview with representatives from the Swedish Transport Agency revealed that the authority does not take into account other countries' fees in the impact statement. The cost variations between countries depend on several factors, including the way rules are interpreted, the financing of the management, and the different divisions between permitting and regulatory activities. It emerged in a dialogue with various respondents that the awareness of fees and rules outside Sweden is limited, since very few respondents emphasized good examples of charging or regulations from other countries.

Respondents want tariff differences, which prevent undesirable behaviour, for example when imposing congestion charges. This has the potential to create a better allocation and use of the infrastructure. According to respondents, some economic instruments that could be differentiated and contribute to better governance include:

- Energy and carbon dioxide taxes on fuel;
- Road pricing for heavy vehicles;
- Infrastructure fees for the railway;
- Vehicle taxes;
- Congestion charges.

Track access charges have already partly been differentiated and a certain degree of differentiation of fairway dues has been introduced based on environmental factors. According to some respondents, further differentiation of fairway dues should be considered primarily to achieve a shift from road and rail transport towards shipping. Suggested changes should come from the relevant authorities for stakeholders and affected users.

Some respondents emphasize that the pricing of carbon emissions should be consistent regardless of where in the transport system the discharge takes place and that this is particularly important for the internalization of external costs for all transport modes. Some respondents wish for the instruments to be adapted in a way that benefits the environment without having a negative impact on the market.

In the discussions with the respondents it was emphasized that the differentiation and internalization of fees should be based on actual costs and equal treatment modes in the pricing, which should also take into account the risk of accidents. Differentiated charging based on external costs can create more efficiency in the system, according to respondents. Further, it was pointed out that the mode of transport that is most effective from a socio-economic perspective should be used to a greater extent.

Some political stakeholders state that a distance driven-based tax should be introduced and differentiated by railway lines and geography. This means that road transports will be more expensive where there are railway tracks available, while at the same time it should be cheaper to run on low-traffic routes and during daytimes with lower utilization. Some respondents underscore the need for more and better "green cars" and that there should be stronger incentives to maintain high standards of rolling stock, particularly in the railway sector.

The rail traffic is charged by the marginal cost principle based on gained paths. In addition to this fee, one of the respondents suggested introducing an additional variable fee that may be charged if the path involves the operation of particularly congested routes. This type of extra fee can therefore be used as an instrument in planning train paths.

For road vehicles, vehicle taxes are levied instead of fees, as for railway traffic. Charges such as the vehicle tax are not a direct incentive for an efficient use of the vehicles. Instead, according to several respondents, the charges imposed should to a greater extent be linked to the use of the means of transport, such as distance-based charges. For calculating the taxes, a more sophisticated system that monitors and records the trucks' mileage is required.

Several respondents call for a clear system/framework for what fees should be charged, how they should be levied, the long-term direction and goals with clear milestones. Moreover, a better system for monitoring and control is needed. Shortcomings in this area can lead to a number of shortcomings and problems in the transport system, such as unfair competition and an increased risk of accidents.

Fees for funding the infrastructure

Fees which are intended, for example, to fund improvements in infrastructure are in principle acceptable, says the respondent from the transport industry organisation Transport Gruppen. At the same time, it was emphasized that it is important to compare the total load on commercial transports all the time in relation to the industry's international competitiveness.

In order to create a willingness to pay for these types of fees, it is important that users have an understanding of what they are paying for. Users also need to know whom they can turn to in order to express their views about the lack of quality of the infrastructure, and a clear link between alleged shortcomings and efforts to improve them is important.

According to respondents, a combination of new, well-designed user fees and reductions in other charges or easing of regulations is an interesting area which should be evaluated further.

Quality charges

One view is that the need for better control and monitoring of who caused which damages and the cost recovery for such damages should be addressed. This is necessary so that fees can be charged directly for the damage incurred. However, it can be difficult to subsequently derive and prove who caused a certain injury. One option is to have a more proactive charging system which attempts to prevent injuries and disruptions. One possible way, according to several respondents, would be to impose fees based on factors such as type of vehicle, engine, etc., to create incentives for the renewal of the vehicle fleet, for example by replacing older vehicles. It was pointed out that today there are quality charges to prevent interference, and that this should be further developed. Rail transports in particular were emphasized as an area with great potential for quality contributions.

One respondent suggested that a premium on transport vehicles which cause less wear on the infrastructure should be implemented, rather than adding additional fees. In this context it was also pointed out that rail operators and authorities providing infrastructure should take greater financial responsibility than they do today for injuries resulting from defective materials.

Instruments for increased utilization of infrastructure

Several respondents emphasized the importance of more and better economic steering instruments and that the capacity assignment should be multimodal and with a greater spread over the day, for example, by using more transports at night-time. Another aspect that was emphasized and demanded is a gradual capacity allocation, which several respondents are aware, is being discussed in various forums. It was also mentioned that a combination of knowledge and legal instruments is necessary. Knowledge-based instruments include information and advice, research and development. Demonstrations and examples of good practice, benchmarking and rating, as well as eco-labelling/certification and environmental management systems are other examples of knowledge-based instruments. Legal instruments include fuel classification/regulation and environmental classification of vehicles. Such instruments also regulate vehicle size, length, capacity and weight, as well as environmental zones and restrictions on vehicle circulation/idling.

An efficient infrastructure requires more testing and evaluation, more maintenance and a more efficient use of infrastructure through a better allocation of capacity. Some respondents believe that new investments are required in order to develop the existing transportation infrastructure, while others believe that there should instead be reallocation and a more efficient utilization of the infrastructure. Several highlight the possibilities of shifting goods transports to shipping.

A number of respondents state that the focus should be more on the 4-step principle, particularly stages 1 and 2. One point raised is that environmental and climate change should be taken into account in a better way. Generally, measures for steps 1 and 2 are more difficult to assess quantitatively than is investment in new infrastructure.

CONCLUSIONS

Based on the major findings from the stakeholder analysis and the comparison with other European countries through earlier studies, the most relevant best practices and benchmarking methods for Swedish conditions could be identified, revealing important interconnections between rule frameworks in Sweden, other European countries, and the EU.

Rule simplification requires that appropriate actions are determined based on business needs through a coordinated simplification work between companies and business enterprises. The Transport Administration conducts regulatory simplification through corrections and additions with new regulations and monitoring. The aim is to simplify, clarify and create more predictable rules while the review of the applications, manuals and forms facilitates contact between businesses and government. Small infrastructure managers are no longer required to issue a network statement and to collect fees. This is expected to cut administrative costs. In addition, the requirements of permits have been abolished for small railway companies with their own goods and infrastructure and/or who only move carts with two-way vehicles.

Joint international rules, rather than national rules, are expected to yield fewer and more harmonized rules and increased efficiency. The ambition is that regulation in the EU should be simpler and clearer through collaboration with government and other EU bodies. The Railway Administration should be able to assist companies in improving the regulation of permits and approvals.

The distribution of responsibility in the public sector should be reviewed and improved in order to create a more homogeneous distribution between the public and private management and financing [16] . System responsibility should be allocated by interest principle, up to the level where it does not give rise to distortions of competition. One possible measure would be to coordinate a Nordic transport department organization for the land based infrastructure in order to create economies of scale and integrate Nordic projects.

New and more private investments could be created by construction engagement forms in public organizations within investment and maintenance. Moreover, one could introduce different organizations for each mainline/corridor and possibly make them responsible for traffic and tracks like in the UK. New railway lines could be separated into various organizations and away from the Swedish Transport Administration for a clearer division of responsibility.

The Swedish regulations should better promote competition and the implementation of alternative forms of financing. In the future, the Swedish Transport Administration should consider solely holding a planning and financing function. Future transport infrastructure should meet a more socio-economic and organizational/business efficiency. The control of appropriated funds can be improved for road and rail networks by extending the appropriation time limits for infrastructure measures. Socio-economic calculations should support the decision basis to allow for better investment planning.

An efficient transport system requires a well-functioning payment system for traffic. Traffic should pay for the costs that it generates. A restructuring of taxes moving from fixed annual fees to variable taxes could free up space and provide opportunities to increase the internalizing taxes. This could contribute to more efficient transportation in the form of minimized travel distances, increased capacity utilization and better road choices.

Congestion problems have a spatial and temporal anchoring and therefore cannot be solved with the general policy instruments. For this reason, they should be analyzed and dealt with by using a specific and adapted instrument that has similar spatial and temporal variations.

The taxes and fees would need to be differentiated according to variables such as vehicle type, geographical location and traffic environment. Differentiation of taxes, however, requires considerably more knowledge about the different types of causal and economic relationships. Today's economic transport statistics are very limited and are based on different types of spatial data. As a result, the transport economics need to develop a better production of economical transport statistics [17] .

Freight transport movements can be promoted by small measures such as simpler waterway improvements, signal systems, interchanges and passing loops. Key initiatives in freight transport infrastructure should focus on port and terminal connections, partial double tracks for railway tracks, buoyancy investment and dredging of waterways. An entire cargo-specific network is required for the transfer of more freight to the rail system, especially in areas where the numerous different flows of the mining and forestry industries meet.

Based on the stakeholder analysis and the comparison with other European countries, this study points to some potential improvements and investigations worth considering from a transport policy perspective. Some examples include:

- Involving industry participants in a review of the capacity allocation process during the analysis of capacity-enhancing measures;
- Monitoring the effects of increased competition on the tracks and their relation to procured services;

- Analyzing the implications of agreements on co-financing for capacity commitments;
- Investigating the requirement for operators to cooperate on a common portal for information;
- Developing action for better and more efficient freight transport market adjustments;
- Investigating how the management of tracks and terminal facilities should be organized, including the role of Jernhusen;
- Considering whether the Transport Administration should have its own resources for the control of the different projects;
- Investigating how Sweden is affected by EU rule developments and consider how Sweden's operations in rule developments can be strengthened;
- Investigating the accumulation of strategic railway expertise, namely where the skills reside and if a new reformed strategic role should be assigned to any organization;

Hilda Karlsson, Rickard Bergqvist, Lars Brigelius · Considering the clarification of roles and responsibilities of different actors and how their cooperation can be improved.

REFERENCES

1. Trafikverket (2011) Hojda banavgifter och deras effekter i ett trafikslagsovergripande perspektiv. Publikation: 2011: 080.
2. Nash, C., Sansom, T., and Matthews, M. (2001) Concerted Action on Transport Pricing Research Integration (CAPRI). Final Report for Publication, Co-Ordinated by the Institute for Transport Studies, University of Leeds, January 2001.
3. Bryman, A. (2008) Samhallvetenskapliga metoder. Liber, Malmo.
4. Alises, A., Vassallo, J.M. and Guzman, A.F. (2014) Road Freight Transport Decoupling: A Comparative Analysis between the United Kingdom and Spain. Transport Policy, 32, 186-193. http://dx.doi.org/10.1016/j.tranpol.2014.01.013
5. Matthews, B., Evangelinos, C., Johnson, D. and Meunier, D. (2009) Impacts and Incentives of Differentiated Rail Infrastructure Charges in Europe—Focus on Freight. European Transport, 43, 83-112.
6. Nash, C. (2005) Rail Infrastructure Charges in Europe. Journal of Transport Economics and Policy, 39, 259-278.
7. Lindberg, G. (2009) CATRIN (Cost Allocation of Transport Infrastructure cost), Deliverable 12—Conclusions and Recommendations. Funded by Sixth Framework Programme, VTI, Stockholm.
8. Nelldal, B.-L., Troche, G. and Wajsman, J. (2009) Effekter av lastbilsavgifter—Analys av effekter av inforande av avstandsbaserade lastbilsavgifter pa konkurrens och samverkan mellan jarnvag och lastbil. Kungliga Tekniska Hogskolan (KTH), Avdelningen for Trafik & Logistik, KTH Jarnvagsgruppen.
9. BMVIT (2014) Bundesministeriumfür Verkehr, Innovation undTechnologie. http://www.bmvit.gv.at/
10. European Commission (1999) EC Directive 1999/62/EC on the Charging of Heavy Goods Vehicles for the Use of Certain Infrastructures.
11. VTI (2013) Internalization of External Effects in European Freight Corridors. CTS Working Paper 2013: 2.
12. Berglund, C.-M. and Eriksson, R. (2003) Pa vag mot marginalkostnadsprissattning inom sjotransportsektorn. VTI meddelande 956.
13. Sjofartsverket (2002) Sjofartens avgiftsrelevanta marginalkostnader; Utvecklingsarbete under 2002. Redovisning av regeringsuppdrag, 31

December 2002.
14. Cullinane, K. and Bergqvist, R. (2014) Emission Control Areas and Their Impact on Maritime Transport, Transport Research Part D, 28D, 1-5.
15. Banverket (2011) Underlagsrapport. Avgifter i Banverkets Jarnvagsnatsbeskrivning 2011, F10-4086/TR00.
16. Hasselgren, B. (2013) Naringslivets transporter—Hur transportinfrastrukturen bli mer effektiv? KTH, Institutionen for Samhallsplanering och miljo.
17. Trafikanalys (2012) Transportsektorns samhallsekonomiska kostnader—Rapport 2012:3.

CHAPTER 15

Spanish Initiative for the Automation in Urban Transport: AutoMOST

Jesús Murgoitio[1], Rafael Durbán[2], Joshué M. Pérez[1], Antonio García[2], Jose A. Moreno[2], Ray A. Lattarulo[1]

[1]Industry & Transport Division, Tecnalia, Derio, Spain

[2]Grupo Avanza, Madrid, Spain

ABSTRACT

The progressive automation of transport will imply a new paradigm in mobility, which will profoundly affect people, logistics of goods, as well as other sectors dependent on transport. It is precise within this automation where the development of new driving technologies is going to cause a great impact on the mobility of the near future, and that will have an effect on the

Citation: Murgoitio, J. , Durbán, R. , Pérez, J. , García, A. , Moreno, J. and Lattarulo, R. (2018), "Spanish Initiative for the Automation in Urban Transport: AutoMOST". *Journal of Transportation Technologies*, **8**, 1-10. doi: 10.4236/jtts.2018.81001.

economic, natural and social environment. It is therefore a primary issue at the global level, as it is reflected in the work programs of the European Commission in relation to the road transport [1] [2] . Thus, the size impact is caused by the following novelties and advantages: 1) Safety: Accidents reduction caused by human error; 2) Efficiency increase in transportation, both in energy consumption and time; 3) Comfort for users and professionals who will increase their operational availability to execute other more valuable tasks, both for them and enterprises; 4) Social Inclusion: enabling mobility easily for everybody during more time; 5) Accessibility, to get to city centers and other difficult reach places. It should be noted that the economic impact projected for automated driving for the years to come ranges up to ?1 bn in 2030, when estimated global market for automated vehicles is 44 million vehicles, as is reflected in document Automated Driving Roadmap by ERTRAC [3] , European Road Transport Research Advisory Council (http://www.ertrac.org/uploads/documentsearch/id38/ERTRAC_Automated-Driving-2015.pdf). As background that already anticipates these improvements, the Advance Driver Assistance System (ADAs) have already showed the safety increase in the last ten years, but always maintain a leading role for the driver. Related to the efficiency increase, automated driving offers great opportunities for those companies where mobility is a key factor in operating costs, and affects the whole value chain. The project opportunity is consistent with ERTRAC vision, especially in applications focused on the urban environment [4] , where it is expected a deployment of the technology of high level automation in an immediate future. This is possible by the potential to incorporate smart infrastructure to improve guidance and positioning, as well as lower speed, which eases its progressive deployment. The objective of AutoMOST is developing technologies for the automation of vehicles in urban transport and industrial applications, to increase significantly the efficiency, safety and environmental sustainability. More specifically, AutoMOST will allow the implementation of shared control systems (Dual-Mode) [5] for future automated vehicles that allow the services operate more efficiently and flexibly, in a context of intelligent and connected infrastructures.

Keywords: Urban Transport, Automated Vehicles, Dual Mode

INTRODUCTION

Transport automation will significantly affect people's lives, merchandise transport and other transport-related sectors [6] [7] . It is precise in this context of progressive automation that the development of new technologies applied to transport becomes an aspect vital to the competitiveness of companies engaged in mobility, affecting their future positions in the economic, environmental and social areas. For all these reasons, road transport automation currently represents a global priority concern, reflected in the work programs of the European Commission [1] [2] concerned with research into land transport. In more concise terms, the benefits expected from transport automation are described below:

- Safety: The reduction of accidents associated to human error.
- Increased transport efficiency (energy consumption, etc.).
- Comfort for users and professionals, increasing resource availability for other tasks.
- Social inclusion, making mobility for everybody possible for longer periods.
- Accessibility in reaching city centres and other areas of difficult access.

With regard to increased efficiency, vehicle automation offers great opportunities for those companies, to whom mobility is critical to operating costs, affecting the entire value chain [7] . In this sense, not only private cars will benefit from the advantages of automation; passenger transport companies, logistics, and other sectors of the urban setting will also benefit from the boon that automated driving offers.

Specifically, AutoMOST, which main information is shown in Figure 1 (see more detailed Figure 2 & Figure 3 too), will enable the implementation of shared (dual mode) control systems for future automated vehicles that will allow for the more efficient and flexible operation of services in a context of interconnected smart infrastructures.

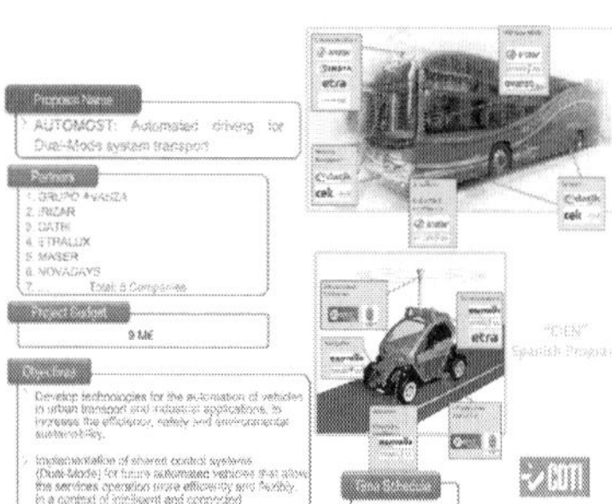

Figure 1. Summarized project data.

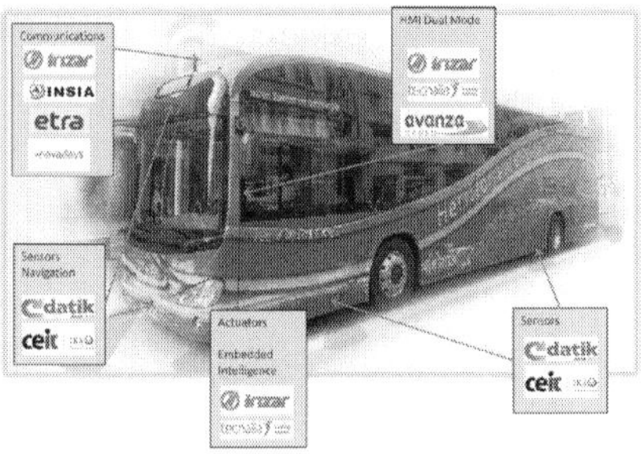

Figure 2. Partner contributions in use case 1: electric bus.

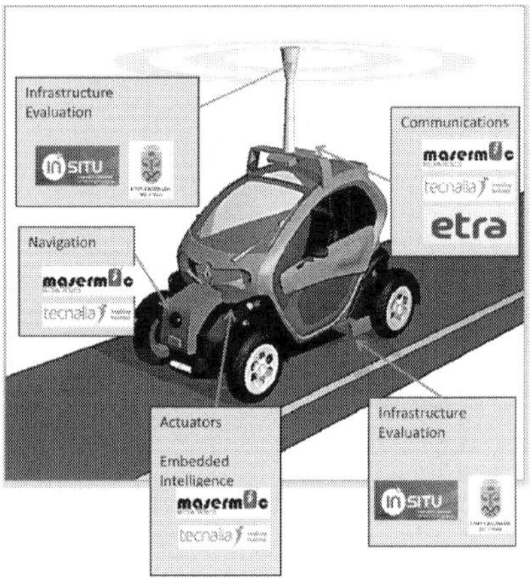

Figure 3. Partner contributions in use case 2: infrastructure maintenance vehicle.

AutoMOST approaches the research and development of enabling elements holistically, simultaneously affecting the value chain to ensure high industrial impact and global leadership positioning for the companies comprising the consortium. In doing this, it departs from the application of existing functions, such as the following driving assistance systems [8] : Adaptive Cruise Control (ACC), Lane Keeping System (LKS), pedestrian detection, smart infrastructures... to facilitate highly efficient operating systems for companies and infrastructures providing mobility, while demonstrating operational cost reductions in specific applications.

Figure 4 shows the advantage of project opportunity, according to a recent study by Morgan Stanley (Company Data, Morgan Stanley Research), which is consistent with ERTRAC vision, especially in applications focused on the urban environment [4] .

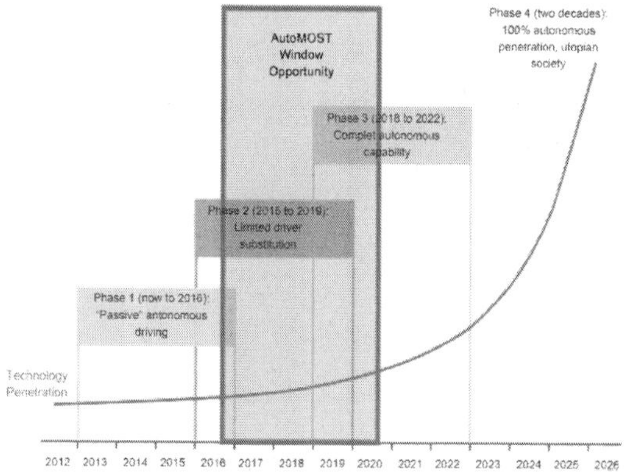

Figure 4. Timeline for technology adoption. Source: Company data, Morgan Stanley research.

APPROACH

The general purpose of AutoMOST is to develop technologies enabling vehicle automation for application to urban and industrial transport with a view to significantly increasing efficiency, safety and sustainability [9] .

Scientific and Technological Objectives

Following are the scientific and technological project objectives arising from the aforesaid general purpose.

- To design and implement a control architecture for automated vehicles of different types (passenger car/large bus), defining each of the blocks in this architecture in a modular way so as to be easily interchangeable, and implementing the different advances in keeping with the new electric/electronic vehicle architectures (in line with standards).
- To implement on-board perception systems for automated vehicles in urban and inter-urban settings, as well as to design controllers for steering (lateral driving) and for pedals (longitudinal driving) that can be used individually and jointly.
- To implement a global and local route generator for real-time vehicle path definition and optimum velocity profile definition based on the data supplied by 3D maps, communication systems, vehicle models and propulsion system status.

- To implement a communications system between vehicles and infrastructures (V2I) that will enable optimizing previously defined profiles and routes, along with the development of infrastructure systems (control centre) for operating the system safely, in tandem with traffic management.

Technical Approach and Structure of Activities

Over the last few years, automated vehicles have been accepted by a great proportion of car manufacturers [7] . The benefits this can bring to transport, along with infrastructure, social, safety [9] , emission and other advantages, place us squarely on the path to one of the ground-breaking technologies of the 21st century.

Technical Approach

AutoMOST focuses on the use of many of the functions already existing for automated vehicles up to the present, avoiding additional efforts with regard to systems already tested and validated. AutoMOST seeks to go beyond by developing technologies that enable automated "dual mode" driving under real traffic conditions, interacting safely with the dynamic urban environment and with expert drivers to validate interaction between different automation levels [3] and thus allow for safe, efficient deployment.

The control architecture to be developed in AutoMOST enables implementation of a modular solution for automated vehicles considering safe behavior in different scenarios, particularly in settings structured at urban speeds (ports, urban and inter-urban areas). Hence, the AutoMOST work methodology is based on the architecture shown in Figure 5 which is depicted below.

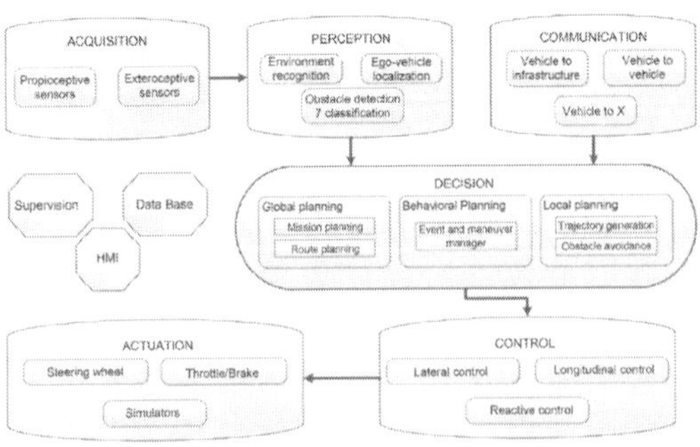

Figure 5. General abstract of automated vehicle control architecture.

This architecture is based on the different advances developed by different international research groups and centres: California PATH, Karlsruhe Institute of Technology (KIT), the University of Parma (and Vislab), Carnegie Mellon University (CMU), Darpa Grand Challenge, Hanyang University, INRIA, etc.

Architectural modularity is a key aspect of the AutoMOST project, since it ensures the adaptability of the system to different computational environments. On another side, the greatest contribution of this architecture is reflected in the dynamic route generator, which takes the conditions of vehicle, infrastructure and obstacles into consideration in defining the route to follow and alternative routes in real time: This block represents the intelligence of the vehicle, which will allow it to negotiate different environments.

To date, most of the contributions to autonomous driving have focused on the modules for perception, communication and activity. AutoMOST focuses on the development of the modules shown in the following figure, with particular emphasis on the decision-making module, in which on-board intelligence will enable more solid control of the different situations in urban environments. The figure below shows the distribution of contributions from each partner, as per the principal on-board vehicle modules. Figure 2 and Figure 3 shows the distribution of contributions from each partner, as per the principal on-board vehicle modules.

Structure of Activities

Figure 6 describes the principal technical activities to be conducted in the AutoMOST project in accordance with the following activity diagram.

- Activity 1: Definition of the solutions presented in the project, in particular scientific and technological objective 1, related to implementing a control architecture for different types of automated vehicle (buses in urban settings and vehicles for infrastructure surveillance).
- Activity 2: Conduct of all tasks related to environmental recognition, mainly along three lines: positioning, obstacle detection and awareness of infrastructure status.

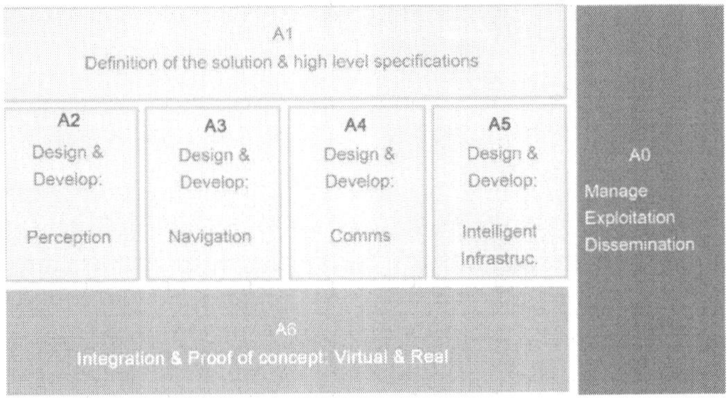

Figure 6. AutoMOST activity diagram.

- Activity 3: Conduct of the tasks related to on-board intelligence: design of direction controllers (lateral driving) and pedal controllers (longitudinal driving), and the implementation of a global route and local path generator for the real-time definition of routes to follow, using the data from the positioning and obstacle detection system and the definition of optimum velocity profiles based on data from 3D maps, communication systems, vehicle models and propulsion system status as reference.
- Activity 4: Implementation of a communications system between the vehicles and the infrastructure (V2I) making it possible to optimize previously defined profiles and routes and warn about some emergency situations.
- Activity 5: Development of infrastructure systems with a central control center for safely operating the automated vehicle system while managing overall city traffic.
- Activity 6: Implementation and testing of the subsystems developed in activities A2-A5 in all use cases: Automated electric bus for passenger transport at the Málaga port and smart vehicles for infrastructure maintenance.

The diagram in Figure 7 shows the timeline for the technical activities projected, including the most relevant milestones.

RESOLUTION

The AutoMOST project is resolved to develop technologies enabling vehicle automation for application to urban transport and thoroughfare status monitoring, with a view to significantly increasing efficiency, safety and sustainability. It is for this reason that a high degree of innovation may be expected as a result of the project. The following are therefore outstanding among other innovations:

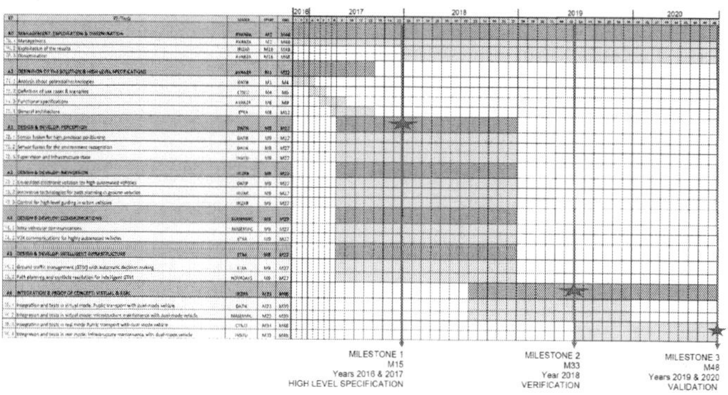

Figure 7. AutoMOST project gantt diagram.

- Integration of the results from different disciplines to automate an electric bus for transport in the urban setting, as well as a vehicle for thoroughfare infrastructure monitoring.
- Development to implement the blocks of a modular control architecture for autonomous vehicles.
- Development to implement a multi-sensor perception solution for autonomous vehicle positioning in urban settings.
- Development of an on-board system operating in real time to generate curve parameters based on data from on-board perception systems.
- Development of a global route generator based on the data from on-board communication systems and sensors.
- Development of a dual system that will enable drivers to choose the type of activators to use in driving, as it is carried out for platoons [10] [11] .

CONCLUSIONS

The challenges identified in research on both navigation and driverless vehicle control systems in dynamic urban environments are at a relatively low stage of maturity. Hence, up to now, many of the capabilities for avoiding accidents with potential obstacles and facilitating cooperative manoeuvring between automated vehicles have been limited by the scope of development projects, which usually resolve very specific scenarios under highly controlled circumstances [12] , making deployment hard in complex environments, where multiple circumstances exist. This makes a decided effort for implementation in real urban settings necessary [4] - [13] .

It may thus be concluded that AutoMOST focuses on the use of many of the functions already existing for automated vehicles up to the present, avoiding additional efforts with regard to systems already tested and validated. AutoMOST, however, seeks to go beyond by developing technologies that enable automated and "dual mode" driving under real traffic conditions, interacting safely with the dynamic urban environment and with expert drivers to validate interaction between different automation levels and thus allow for safe and efficient deployment.

ACKNOWLEDGEMENTS

This work is supported by the Centre for the Development of Industrial Technology (CDTI) through the CIEN (Consorcios de Investigación Empresarial Nacional) Spanish program. So, the authors would like to thank the CDTI which is a Public Business Entity, answering to the Ministry of Economy and Competitiveness in Spain, which fosters the technological development and innovation of Spanish companies. It is the entity that channels the funding and support applications for national and international R&D&I projects of Spanish companies.

REFERENCES

1. ERTRAC: European Road Traffic Research Advisory Council (2013) Multi-Annual Implementation Plan for Horizon 2020, March 2013. http://www.ertrac.org/
2. European Commission, Roadmap to a Single European Transport Area (2011) Towards a Competitive and Resource Efficient Transport System, 2011.
3. ERTRAC: European Road Traffic Research Advisory Council (2015) Automated Driving Roadmap, 2015. http://www.ertrac.org/
4. Mitchell, W.J., Borroni-Bird, C.E. and Burns, L.D. (2010) Reinventing the Automobile: Personal Urban Mobility for the 21st Century. MIT Press, Cambridge, MA, 225 p.
5. Larburu, M. and Sanchez, J. (2010) Safe Road Trains for Environment: Human Factors' Aspects in Dual Mode Transport Systems. 17th World Congress on Intelligent Transport Systems, Busan, Korea, 25-29 October 2010.
6. Mohanarangam, K., et al. (2014) Strategic Insight into Voice of European Consumers on Passenger Car Safety Systems, Desirability and Willingness to Pay for Active Safety and ADAS Features, Frost & Sullivan.
7. Natarajan, P., Jayaraman, K., Chandrasekar, P. and Ford, N. (2014) Strategic Analysis of the European and North American Market for Automated Driving Frost & Sullivan.
8. SAE (2014) International On-Road Automated Vehicle Standards Committee Taxonomy and Definitions for Terms Related to On-Road Motor Vehicle Automated Driving Systems SAE International.
9. ERTRAC: European Road Traffic Research Advisory Council (2011) Working Group on Road Transport Safety and Security, European Roadmap on Safe Road Transport. http://www.ertrac.org/
10. Solyom, S. and Coelingh, E. (2012) Performance Limitations in Vehicle Platoon Control. Proceedings of the 15th IEEE Intelligent Transportation Systems Conference, Anchorage, US, 16-19 September 2012. https://doi.org/10.1109/ITSC.2012.6338689
11. Robinson, T., Chan, E. and Coelingh, E. (2010) Operating Platoons on Public Motorways: An Introduction to the SARTRE Platooning Programme. 17th World Congress on Intelligent Transport Systems, Busan, Korea, 25-29 October 2010.

12. ADAM: Automation Development for Autonomous Mobility (2014) Iniciativa financiada por el programa INNPRONTA, 2011-2014.
13. Parent, D. (2011) CYBERCARS: New Developments and Future Applications. Trans-portation Research Board 88th Annual Meeting.

CHAPTER
16

Fuzzy Analysis and Evaluation of Public Transport Service Quality: A Case Study of Dar es Salaam City, Tanzania

Erick P. Massami[1], Benitha M. Myamba[2], Leticia Edward[3]

[1]Unit for Applied Mathematics, Dar es Salaam Institute of Technology, Dar es Salaam, Tanzania

[2]Department of Logistics and Transport Studies, National Institute of Transport, Dar es Salaam, Tanzania

[3]Department of Computing and Communication Technologies, National Institute of Transport, Dar es Salaam, Tanzania

ABSTRACT

Urban public transport plays a critical role in stimulating economic development of any nation since most of the revenues come from cities. The majority of the city dwellers of any country use public transport. The

Citation: Massami, E. , Myamba, B. and Edward, L. (2016), "Fuzzy Analysis and Evaluation of Public Transport Service Quality: A Case Study of Dar es Salaam City, Tanzania". *Journal of Transportation Technologies*, 6, 297-311. doi: 10.4236/jtts.2016.65027.

evaluation of public transport service quality provides a valuable feedback to commuter operators to ensure continuous improvement of level of service and to the government to take appropriate measures for enhancing the quality of public transport service. This paper analyses and evaluates service quality of Road Public Transport (RPT) (i.e. minibuses and buses) and Urban Rail Transport (URT) in Dar es Salaam City, Tanzania. Since service quality and its attributes are intangible and vague, a fuzzy evaluation model is developed and applied. The formulated model is composed of Fuzzy Entropy Method (FEM) and Technique for Order Preference by Similarity to Ideal Solution (TOPSIS). The overall evaluation procedure is as follows: initially, an intensive literature search and experts' opinions are employed to establish criteria for evaluating the service quality of public transport in Dar es Salaam City. The developed FEM is used to obtain criteria weight. Lastly, the formulated TOPSIS is used to provide an overall ranking of urban public transport service quality. The overall evaluation shows that urban rail transport outperforms road public transport in terms of service quality. Nevertheless, the urban rail transport service in Dar es Salaam City is currently not well developed as it is provided on very limited routes. Thus, the Tanzania government, the rail transport operator i.e. Tanzania Railway Limited (TRL) and the agency responsible for the provision of rail infrastructure i.e. Reli Assets Holding Company (RAHCO) are advised to design and employ Public-Private-Partnership (PPP) schemes i.e. concession contracts to invest more in rolling stocks, locomotives and rail wagons so that the rail transport service is available on many routes of the road public transport to bring fair competition between the two operators.

Keywords: Service Quality, Urban Public Transport, Entropy Method, Fuzzy TOPSIS, Dar es Salaam City

INTRODUCTION

Public transport plays a critical role in stimulating economic development of any nation. Majority of the population of any country use public transport. Transport differs from other challenges developing societies face as it gets worse rather better with economic development. While sanitation, education, and other challenges improve with economic growth, transport gets worse. An efficient transport system is crucial to a nation's development and can have far reaching benefits, and conversely, failing to do so will have an adverse repercussion on development. The rate of population growth in most developing countries is higher than the rate of car ownership, and therefore an increasing proportion of the population depends on public transport. Public transport also provides an alternative to private car owners who sometimes are deterred from using it by traffic congestion, parking difficulties or difficulties in accessing certain streets. In addition, most families of developing countries tend to be large, so that even if there is a car belonging to a household, most members of the family are still likely to use public transport for at least some of their travel. Thus, it is vital for the industry transport stakeholders to analyse and/or evaluate the public transport service quality to improve the standard of life. More specifically, evaluating the transport service level helps commuter operators understand their overall service quality in terms of manageable service attributes, thus identifying service areas to be improved with the index service as a service benchmarking and management tool. As such, service quality level has a very important significance for Transport Company's competitiveness in a free market.

The government is also benefited from the efficient transport system which in turn stimulates the country's development. Nonetheless, most of the cities worldwide have been experiencing tremendous transport challenges including inadequacy of proper public transit service. In particular, Msigwa [1] identifies the challenges facing transportation in the cities of Tanzania. Massami and Myamba [2] evaluate the challenges facing public transportation in Dar es Salaam City, Tanzania in which the inadequacy of proper public transit service is identified as the second challenge.

Dar es Salaam City, being a Tanzanian financial and economic centre, plays a very big role for the development of the country as whole. In particular, Dar es Salaam sea port and the Dar es Salaam based Julius Nyerere International Airport (JINIA) are the main nodes for freight and passengers' supply chain in the international trade for Tanzanians and residents of the landlocked countries of Uganda, Rwanda, Burundi, East Democratic Republic of Congo, Zambia and Malawi. Thus, it is vital to measure the level of public transport service quality in Dar es Salaam City. Nonetheless, the evaluation of public transport service quality in the developing countries (e.g. Tanzania) is hardly investigated by the public transport operators and/or the government.

Traditionally, academicians and practitioners use SERVQUAL and SERVPERF for service quality evaluation. Although these studies address the ways of measuring the attributes' effectiveness, the imprecise nature of passenger's perception on service quality is not addressed [3] .

Since service or level of service is considered an imprecise quantity, many researchers and practitioners are currently proposing models based on Fuzzy Set Theory (FST).

The paper is structured as follows: Section 2 presents a survey of previous work; modelling framework for urban public transport service quality is presented in Section 3; evaluation of public transport service quality in Dar es Salaam City is given in Section 4 and conclusions are presented in Section 5.

SURVEY OF PREVIOUS WORK

In recent years, service quality has become one of the most important issues in transport management [4] . Oña et al. [5] apply a cluster analysis technique for stratifying the sample of users of a public transport service in the city of Granada (Spain) in order to analyze passengers' opinions under more homogeneous conditions. Stelzer et al. [6] use a novel approach which uses standardised automated information exchange between travellers and Transportation Company tailored to improve the service quality of public transport. Mouwen [7] proposes a study to improve the understanding of the drivers of customer satisfaction with public transport (PT). D'Ovidio et al. [8] apply factor analysis with categorical principal component analysis and multivariate methods of classification to explore the characteristics of the customers and their quality perception on bus service of the town of Bari. Wang et al. [9] apply compound Poisson processes to formulate queues of batch arrivals and bulk service including balking and reneging behaviours of customers. Nicolás et al. [10] present a study aimed at providing the municipality of Alcoy-Spain with detailed information about public transport service based on functional, economic, social and spatial coverage. Bruglieri et al. [11] propose a study which focuses on the design and development of a real time mobility information system for the management of unexpected events, delays and service disruptions concerning public transportation in the city of Milan. Olsson et al. [12] measure the service experience in public transport using the Satisfaction with Travel Scale (STS). Khalid et al. [13] analyse data using SPSS to investigate user perceptions of rail public transport services (KTM Komuter) in Kuala Lumpur, Malaysia.

Song et al. [14] develop quantitative indicators for public transportation service evaluation and utilize data mining technique which derives data from the smart card system in the city of Seoul. Isabello et al. [15] review efficiency and effectiveness of interurban public transport services of the

Piedmont region of Italy. Redman et al. [16] use a qualitative systematic review to better understand those aspects of public transport quality which most likely attract car users. Olio et al. [17] propose the methodology which provides the authorities and operating companies with useful information to plan personalised marketing policies at different categories of users and potential users of public transport. Daraio et al. [18] address the problem of efficiency and effectiveness of local public transport (LPT) with special emphasis on urban transit. Yaliniz et al. [19] evaluate the transit quality of service in Kutahya, Turkey based on aspects such as transit availability, comfort and convenience. Anna et al. [20] propose a panel data approach to analyze the passenger satisfaction of a public transport service. Lai and Chen [21] use passenger survey data from the Kaohsiung Mass Rapid Transit (KMRT)— Taiwan and apply the structural equation modelling to explore the relationships between passenger behavioural intentions and the factors that affect them. Oña et al. [22] use data from a customer satisfaction survey collected by the transport consortium of Granada (Spain) apply a Structural Equation Model (SEM) to reveal the unobserved latent aspects describing bus transit service quality. Del Castillo and Benitez [23] apply models based on averages, multivariate discrete distribution and a generalized linear model to identify the aspects that mostly influence the perception of overall transport service quality.

In contrast to the most surveyed literature that employs statistical methods, very few studies employ soft computing techniques to evaluate transport service quality. Garrido et al. [24] propose Artificial Neural Networks (ANN) to analyze the service quality perceived by the passengers of Granada bus metropolitan transit system. Nonetheless, little attention is paid to evaluation of service quality in the Tanzanian urban transport environment. In addition, the evaluation of transport service quality is characterized by uncertainty, subjectivity, imprecision and ambiguity. Fuzzy set theory has been proven to be a useful mathematical tool for modelling uncertain i.e. imprecise and vague data in real situations [25] .

Nowadays there are heaps of articles on the theory of fuzzy entropy and its application in weight determination. For instance, Gong et al. [26] use fuzzy entropy approach to determine the weights used to evaluate the reliability of emergency logistics system. Bray et al. [27] develop feature selection method based on fuzzy entropy and apply to a set of international container ports which is considered as Decision Making Units (DMUs). Lin and Tan [28] apply fuzzy entropy for weight determination in the evaluation of port development. On the other hand, fuzzy TOPSIS has been extensively applied for criteria assessment and ranking of alternatives in various fields. Behzadian et al. [29] provide useful insights of a state-of the-art survey of TOPSIS applications including supply chain management and logistics; design, engineering and manufacturing systems; business and marketing management, health, safety and environment management; human resources management; energy management; chemical engineering; water resources management and other topics. Awasthi et al. [30] propose a hybrid approach based on SERVQUAL and fuzzy TOPSIS for evaluating service quality of metro in Montreal. Torlak et al. [31] use fuzzy TOPSIS approach to evaluate business competition in the Turkish domestic airline industry. Awasthi et al. [32] present a multi-criteria decision making approach for selecting sustainable transportation systems under partial or incomplete information (uncertainty). Toosi and Kohanali [33] use customer survey and apply AHP and fuzzy TOPSIS to evaluate service quality of three airlines in Qeshm free zone in Iran.

This study therefore presents a composite model of fuzzy entropy and fuzzy TOPSIS applied to evaluate transport service quality in Dar es Salaam City, Tanzania. The proposed approach has proven to be a powerful modelling tool for coping with subjectivity and imprecision in human judgements associated with service quality evaluation. The proposed composite model can provide transport operators and/or the government with valuable inputs on the analysis and evaluation of the level of transport service.

MODELLING FRAMEWORK FOR URBAN PUBLIC TRANSPORT SERVICE QUALITY

The fuzzy multi-criteria decision approach presented in this study is composed of two stages. The first stage describes the steps to obtain the evaluation performance for each criterion of the service quality while the second stage describes the steps to obtain the overall service quality of public transport.

Evaluation of Performance Criteria

Researchers and practitioners have proposed a number of methods for calculating the performance of criteria. These approaches are such as Analytical Hierarchy Process (AHP), Triangular Fuzzy Number (TFN), Fuzzy comparison method etc. In this study, Fuzzy Entropy Method (FEM) is used to obtain the performance of criteria. The FEM is a weight determination approach which gives the weight values of individual criteria and/or sub-criteria by calculating the entropy and entropy weight [34] . The larger the entropy, the smaller its entropy weight. When the entropy weight is zero, no useful information is provided to the decision-maker, and this criterion and/or sub-criterion can be removed from the process [35] .

The following are the main steps of the FEM.

Formation of the Standardized Evaluation Matrix

The fuzzy evaluation matrix, $V=\{v_j\},\ j\in\{1,2,\cdots,J\}$ where $r_{ij}\in[0,1]$ is the membership grade of the ith criterion/sub-criterion to the jth evaluation level. This matrix R is a result of accomplishment of the following steps.

1) Formation of the Evaluation Set

An evaluation set $V=\{v_j\},\ j\in\{1,2,\cdots,J\}$ and v_j stands for the jth evaluation level. This study proposes the evaluation set $V=\{\text{Very dissatified, Not satisfied, Fair, Satisfied, Very satisfied}\}$.

2) Determination of the Membership Grade

The evaluators' scores are used to determine the membership grade of the evaluation set on each criterion. For instance, suppose there are N experts (i.e. evaluators) in which $x_{ij}\ (j=1,2,3,\cdots,J)$ experts believe that the influence of the criterion i (e.g.

convenience) belongs to the evaluation level $v_j,\ j\in\{1,2,\cdots,J\}$, then $r_{ij}=\dfrac{x_{ij}}{\sum_{j=1}^{J}x_{ij}}$ is the evaluation score. Thus, the single-factor evaluation matrix can be expressed by Equation (1):

$$R=\left[\frac{x_{ij}}{\sum_{j=1}^{J}x_{ij}}\right],i\in\{1,2,3,\cdots,I\},j\in\{1,2,\cdots,J\} \tag{1}$$

Weight Determination

Suppose we have the standardized evaluation matrix $R=\left[r_{ij}\right]_{IJ}$ Now the entropy of the ith criterion is defined by Equation (2):

$$H_{mi}=-(\ln J)^{-1}\sum_{j=1}^{J}r_{ij}\ln(r_{ij}),m\in\{1,2,3,\cdots,M\},i\in\{1,2,3,\cdots,I\} \tag{2}$$

The weight of entropy of the ith criterion corresponding to alternative m is defined by Equation (3):

$$W_{mi} = \frac{1 - H_{mi}}{I - \sum_{i=1}^{I} H_{mi}}, m \in \{1,2,3,\cdots,M\}, i \in \{1,2,3,\cdots,I\} \quad (3)$$

where $0 \leq W_{mi} \leq 1$ and $\sum_{i=1}^{I} W_{mi} = 1$.

Fuzzy TOPSIS

The TOPSIS method is proposed by Hwang and Yoon [36] . The fuzzy TOPSIS has proven to be an effective approach for ranking a finite number of alternatives characterized by multiple criteria [37] . This method defines the positive ideal solution (PIS) and the negative ideal solution (NIS). The PIS is the solution which gives the maximum benefit alternative and minimum cost alternative. Meanwhile the NIS is the one which gives the maximum cost alternative and minimum benefit alternative. The optimum value is the one which is closest to the PIS and farthest from the NIS. The general procedure for fuzzy TOPSIS starts by establishing the fuzzy decision matrix. This fuzzy decision matrix is used to deduce the normalized i.e. the weighted decision matrix. However, the elements of the weighted decision matrix are already given by the FEM. Consequently, the fuzzy TOPSIS modelling approach employed in this study involves the computation of the following:

The weighted (normalized) fuzzy decision matrix

The weighted fuzzy decision matrix is given by Equation (4)

$$V = [v_{mi}]_{MI}, v_{mi} = w_{mi}, m \in \{1,2,\cdots,M\}, i \in \{1,2,\cdots,I\}. \quad (4)$$

The fuzzy positive ideal solution (FPIS) and fuzzy negative ideal solution (FNIS)

The FPIS of the alternatives is computed by Equation (5):

$$A^{+} = (v_1^{+}, v_2^{+}, \cdots, v_M^{+}), v_m^{+} = \max_m \{v_{mi}\}, m \in \{1,2,\cdots,M\}, i \in \{1,2,\cdots,I\}. \quad (5)$$

The FNIS of the alternatives is computed by Equation (6):

$$A^{-} = (v_1^{-}, v_2^{-}, \cdots, v_M^{-}), v_m^{-} = \min_m \{v_{mi}\}, m \in \{1,2,\cdots,M\}, i \in \{1,2,\cdots,I\}. \quad (6)$$

The distance of each alternative from FPIS and FNIS

The distance d_m^{+} of v_{mi}, $m \in \{1,2,\cdots,M\}$ from the FPIS is computed by Equation (7):

$$d_m^+ = \sum_{i=1}^{I} d\left(v_{mi}, v_m^+\right) = \sqrt{\sum_{i=1}^{I}\left(v_{mi} - v_m^+\right)^2}, m \in \{1, 2, \cdots, M\}. \tag{7}$$

Moreover, the distance d_m^- of v_{mi}, $m \in \{1, 2, \cdots, M\}$ from the FNIS is computed by Equation (8):

$$d_m^- = \sum_{i=1}^{I} d\left(v_{mi}, v_m^-\right) = \sqrt{\sum_{i=1}^{I}\left(v_{mi} - v_m^-\right)^2}, m \in \{1, 2, \cdots, M\} \tag{8}$$

where d(a, b) is the distance between two fuzzy numbers a and b.

The closeness coefficient (CC_m)

The closeness coefficient of each alternative (CC_m) is calculated by Equation (9).

$$CC_m = \frac{d_m^-}{d_m^- + d_m^+}, m \in \{1, 2, \cdots, M\} \tag{9}$$

Where $d_m^- \geq 0$, $d_m^+ \geq 0$, $CC_m \in [0, 1]$. Thus, an alternative m becomes closer to A^+ as CC_m approaches to 1.

Ranking of the alternatives

The alternatives are ranked in decreasing order based on the values of the closeness coefficient (CC_m). Thus, the most performing alternative is closest to the FPIS and farthest from the FNIS.

EVALUATION OF PUBLIC TRANSPORT SERVICE QUALITY IN DAR ES SALAAM CITY

The literature proposes numerous service attributes that influence the choice of public transport service. Currently there are two types of commuter services in Dar es Salaam City: Road Public Transport (RPT) composed of urban buses and mini-buses which operate on all legitimate routes and Urban Rail Transport (URT) which operate on very few routes with rail rolling stocks in place. This study therefore proposes only those attributes that are relevant to both commuter services (i.e. urban rail and urban buses/ min-buses). These service attributes are shown in Table 1.

Data Collection Method

A survey questionnaire was used to collect data from fifty (50) targeted respondents who gave their perceptions on the attributes of transport service quality. The sample of fifty (50) respondents has been considered based on the fact that the proposed fuzzy evaluation approach handles efficiently problems associated with vagueness and small data. In order to get useful inputs, the respondents chosen possessed at least a diploma qualification. We believe this kind of respondents could think critically and give unbiased assessments. In addition, the interviewer clarified the transport service attributes and their components immediately before commencing the assessment exercise. Moreover, the purposive sample composed of respondents who often use both Road Public Transport (RPT) and Urban Rail Transport (URT).

Table 1. Service standards of public transport in Dar es Salaam City.

Code	Service Standard	Performance Measures
C_1	Comfortability	Type of seating, seat spacing, seat design and materials used
		The proportion of seated to standing passengers
		Heating and ventilation systems
		Luggage accommodation
C_2	Security	Pickpocketing on buses/trains and at bus stops and railroad stations
		Good lighting on buses/trains and at bus stops/train stations
		Presence of Conductors/Inspectors on the buses/trains
C_3	Convenience	Accessibility of bus stops/train stations in different parts of the city
		Frequency of service
		Interchangeability between services
C_4	Safety	Driving standards (i.e. Driving speed)
		Bus/train condition
		Overloading
		Road/Railway conditions
C_5	Reliability and Punctuality	Constant operation of a scheduled service
		Operation of service according to scheduled departure and arrival times

C_6	Attendants' Interpersonal Skills	Passenger complaints handling
		Courtesy of attendants
		Helpful attitudes
		Smartness of attendants

The Weights of Evaluation Criteria

After analysing the collected data we get the following:

The standardised evaluation matrix for the road public transport service (R_1) is given by Equation (10):

$$R_1 = \frac{1}{50}\begin{bmatrix} 17 & 22 & 10 & 1 & 0 \\ 25 & 16 & 9 & 0 & 0 \\ 3 & 16 & 20 & 11 & 0 \\ 9 & 29 & 10 & 2 & 0 \\ 4 & 21 & 19 & 6 & 0 \\ 19 & 17 & 11 & 3 & 0 \end{bmatrix}. \quad (10)$$

Thus,

$$R_1 = \begin{bmatrix} 0.34 & 0.44 & 0.20 & 0.02 & 0.00 \\ 0.50 & 0.32 & 0.18 & 0.00 & 0.00 \\ 0.06 & 0.32 & 0.40 & 0.22 & 0.00 \\ 0.18 & 0.58 & 0.20 & 0.04 & 0.00 \\ 0.08 & 0.42 & 0.38 & 0.12 & 0.00 \\ 0.38 & 0.34 & 0.22 & 0.06 & 0.00 \end{bmatrix}. \quad (11)$$

The standardized evaluation matrix for the urban rail transport service (R_R) is given by Equation (12):

$$R_2 = \frac{1}{50}\begin{bmatrix} 2 & 11 & 19 & 12 & 5 \\ 3 & 12 & 27 & 7 & 1 \\ 13 & 21 & 13 & 3 & 0 \\ 1 & 8 & 28 & 12 & 1 \\ 8 & 14 & 16 & 9 & 3 \\ 2 & 16 & 23 & 8 & 1 \end{bmatrix}. \quad (12)$$

Thus,

$$R_2 = \begin{bmatrix} 0.04 & 0.22 & 0.38 & 0.26 & 0.10 \\ 0.06 & 0.24 & 0.54 & 0.14 & 0.02 \\ 0.26 & 0.42 & 0.26 & 0.06 & 0.00 \\ 0.02 & 0.16 & 0.56 & 0.24 & 0.02 \\ 0.16 & 0.28 & 0.32 & 0.18 & 0.06 \\ 0.04 & 0.32 & 0.46 & 0.16 & 0.02 \end{bmatrix}. \quad (13)$$

We apply Equations (2) and (3) to compute the entropy and weight of each criterion for RPT and URT using the elements of the two standardized matrices given by Equations (11) and (13). The perception of passengers on service attributes for each of the transport service category is revealed in the ranking deduced from the calculated entropies and weights as shown in Table 2.

Table 2. Entropy, weight and rank of criteria for RPT and URT.

Criterion (C_i)	H_{1i}	W_{1i}	R_{1i}	H_{2i}	W_{2i}	R_{2i}
C_1	0.701	0.173	3	0.876	0.101	5
C_2	0.634	0.212	1	0.744	0.207	2
C_3	0.766	0.136	5	0.766	0.190	4
C_4	0.668	0.192	2	0.694	0.248	1
C_5	0.738	0.152	4	0.927	0.059	6
C_6	0.768	0.135	6	0.759	0.195	3

Note: H_{1i}: criterion entropy for RPT; H_{2i}: criterion entropy for URT; W_{1i}: criterion weight for RPT; W_{2i}: criterion weight for URT, R_{mi}: rank of criterion i corresponding to alternative m.

From Table 2, it is found that the customers i.e. passengers perceive the service quality of urban road transport in the order given in the set {Security, Safety, Comfortability, Reliability and Punctuality, Convenience, Attendants' interpersonal skills}. Thus, the passengers who use road transport in Dar es Salaam city are less concerned about the security, safety and comfortability aspects of the service and more concerned about the attendants' interpersonal skills, convenience, reliability and punctuality of the service. The findings suggest that the urban road transport operators can attract more passengers provided the following are done: Bus conductors and drivers are well trained and adhere to the professional code of conduct, operate scheduled services and increase the frequency of service.

On the other hand, the passengers perceive the service quality of urban rail transport in the order given in the set {Safety, Security, Attendants' interpersonal skills, Convenience, Comfortability, Reliability and Punctuality}. Thus, the passengers who use urban rail transport are less concerned about the safety and security attributes of the urban rail transport service and more concerned about the reliability and comfortability attributes. These findings require the rail transport operators to make sure there is a constant operation of a scheduled service and increase the frequency of service which in turn would decrease the proportion of seated to standing passengers in order to be more competitive in the public transport market.

Nonetheless, the RPT relatively outperforms the URT on the following attributes of service quality: reliability and punctuality, security and comfortability whereas the URT relatively outperforms the RPT on the following attributes of service quality: safety, attendants' interpersonal skills and convenience.

In order to win the urban public transport market, each operator needs to improve each service attribute over its competitor and formulate effective marketing strategies. The role of the government in this regard should be to formulate attractive policies to the investors and enforce the regulations effectively.

Ranking of Urban Road Transport and Urban Rail Transport

The fuzzy TOPSIS is applied to obtain the overall ranking of transport service quality by computing the following:

The Weighted Fuzzy Decision Matrix

The elements of the weighted fuzzy decision matrix (V) are obtained by applying Equation (4).

$$V = \begin{bmatrix} 0.173 & 0.212 & 0.136 & 0.192 & 0.152 & 0.135 \\ 0.101 & 0.207 & 0.190 & 0.248 & 0.059 & 0.195 \end{bmatrix}.$$

The Fuzzy Positive Ideal Solution (FPIS) and Fuzzy Negative Ideal Solution (FNIS)

Equation (5) is applied to obtain the FPIS:

$$A^{+}=\left(V_1^{+},V_2^{+}\right)=\left(0.212,0.248\right).$$

The fuzzy negative ideal solution (FNIS) is calculated by using Equation (6):

$$A^{-}=\left(V_1^{-},V_2^{-}\right)=\left(0.135,0.059\right).$$

The Distance of the Alternative from the FPIS and FNIS

Equations (7) and (8) are respectively applied to get the distance of the alternative from the FPIS and FNIS as follows:

$$d_1^{+}=\sqrt{\sum_{i=1}^{6}\left(V_{1i}-V_1^{+}\right)^2}=0.131$$

$$d_2^{+}=\sqrt{\sum_{i=1}^{6}\left(V_{2i}-V_2^{+}\right)^2}=0.255$$

$$d_1^{-}=\sqrt{\sum_{i=1}^{6}\left(V_{1i}-V_1^{-}\right)^2}=0.104$$

$$d_2^{-}=\sqrt{\sum_{i=1}^{6}\left(V_{2i}-V_2^{-}\right)^2}=0.308.$$

The Closeness Coefficient (CC_m)

The closeness coefficient for the Road Public Transport (RPT) is given by Equation (9):

$$CC_1=\frac{d_1^{-}}{d_1^{-}+d_1^{+}}=\frac{0.104}{0.104+0.131}=0.443.$$

For the Urban Rail Transport (URT), the closeness coefficient is given by Equation (9):

$$CC_2=\frac{d_2^{-}}{d_2^{-}+d_2^{+}}=\frac{0.308}{0.308+0.255}=0.547.$$

As $CC_2=0.547>CC_1=0.443$, the overall ranking of urban transport service in Dar es Salaam show that the urban rail service offers passengers, the service of relatively high quality. Nevertheless, the service comparison is not involving the newly operated Dar es Salaam Rapid Transit (DART) service which is offered on very limited routes. It should also be noted that the urban rail transport service in Dar es Salaam city is not well developed as it is provided on very few routes. From the analysis the following observations are revealed:

1. The government, the urban road transport and rail transport operators have the obligation of improving the urban transport service to meet the expectation of the current and potential passengers.
2. The rail transport operator, Tanzania Railway Ltd (TRL); the government agency responsible for the provision of rail infrastructure, Reli Assets Holding Company (RAHCO); and the government need to invest more into this transport service segment to meet the current and projected passengers' demand. More specifically, they have to invest more in rolling stocks, locomotives and rail wagons so that the services are available on many routes of the road public transport to bring fair competition between the two operators. As the government is constrained with financial resources, the expansion of the rail transport service can be achieved by involving the private sector through Public-Private Partnerships (PPPs). There are various schemes i.e. concession contracts available in the market such as Build-own-operate-transfer (BOOT), Build- own-operate (BOO), Build-own-transfer (BOT), Equip-operate-transfer (EOT) and Build- transfer-operate (BTO).
3. The overall quality of service offered by the urban transport operators in the Dar es Salaam city has yet met the passengers' expectation. Thus, both urban transport operators should enhance their transport services.
4. The expected transport service quality in the Dar es Salaam city is hardly attained unless; the rail transport operators offer the service on all legitimate routes of the road transport service. This would create fair competition between the two transport operators and curb the envisaged monopolistic practices of the RPT. Nonetheless, Investment in commuter rail transit is a world-wide phenomenon, driven by the belief in its benefits, such as less congestion, less air pollution, improved labour market for the poor, and higher productivity in industries that make substantial use of vehicles in their production processes [38] . More specifically, if the rail transit is developed, it would slow down the rapid increase in auto use which in turn would benefit not only the local residents, but also the rest of the world affected by the massive greenhouse gas emissions from the road transport.

CONCLUSION

In this paper, a composite model of fuzzy entropy and fuzzy TOPSIS is proposed to analyse and evaluate the quality of transport service in Dar es Salaam City, Tanzania. The findings reveal that the overall service quality of urban transport is yet satisfactory i.e. the rail transport service is available during peak hours (morning and evening rush hours) excluding Sundays and public holidays meanwhile the road public transport, with the exception of Dar es Salaam Rapid Transit (DART), has contributed to high levels of congestion which in turn generates economic losses through wasted time, higher travel costs, poor road safety, and high levels of vehicle emissions. The evaluation results reveal that the rail transport service outperforms the road public transport service. In addition, the evaluation procedure generalises the application of fuzzy model to problems involving vagueness. Thus, the results can be an impetus to the transport operators and the government to take needful measures which include minimizing vehicle operating costs and formulating investment policies which will attract the private sector participation through designing various concession contracts. Nonetheless, the Tanzania government has started to implement the PPP scheme by completing the first phase of a six-phase Bus Rapid Transit (BRT) system. Since BRT is operated on very few routes of Dar es Salaam City, the public transit service is very limited. It should be noted that the magnitude of financial assistance needed for urban transport in Dar es Salaam as well as the rest of the country cannot just rely on development partners as the requirements significantly exceed their country allocations. The government needs to examine alternative sources of financing from local sources such as rates and taxes on property and land values. The proposed fuzzy model is recommended as an evaluation tool to other problems involving vagueness and scarcity of data. The future direction of this study is the analysis of the efficiency of the proposed concession contracts as a tool for funding the city's transport system to identify the most suitable PPP schemes in the Tanzanian environment.

ACKNOWLEDGEMENTS

We are very grateful for all respondents who participated in the data collection exercise.

REFERENCES

1. Msigwa, R.E. (2013) Challenges Facing Urban Transportation in Tanzania. Mathematical Theory and Modeling, 3, 18-26.
2. Massami, E.P. and Myamba, B.M. (2016) Application of Vague Analytical Hierarchy Process to Prioritize the Challenges Facing Public Transportation in Dar es Salaam City— Tanzania. International Journal of Advanced Research in Artificial Intelligence, 5, 46-53.
3. Yeh, C.H. and Kuo, Y.L. (2003) Evaluating Passenger Services of Asia-Pacific International Airports. Transportation Research Part E, 39, 35-48. http://dx.doi.org/10.1016/S1366-5545(02)00017-0
4. Lazim, A. and Wahab, N. (2010) A Fuzzy Decision Making Approach in Evaluating Ferry Service Quality. Management Research and Practice, 2, 94-107.
5. Oña, R., L□pez, G., Rios, F.J.D. and Oña, J. (2014) Cluster Analysis for Diminishing Heterogeneous Opinions of Service Quality Public Transport Passengers. Procedia—Social and Behavioral Sciences, 162, 459-466. http://dx.doi.org/10.1016/j.sbspro.2014.12.227
6. Stelzer, A., Englert, F., H□rold, S. and Mayas, C. (2016) Improving Service Quality in Public Transportation Systems Using Automated Customer Feedback. Transportation Research Part E: Logistics and Transportation Review, 89, 259-271. http://dx.doi.org/10.1016/j.tre.2015.05.010
7. Mouwen, A. (2015) Drivers of Customer Satisfaction with Public Transport Services. Transportation Research Part A: Policy and Practice, 78, 1-20. http://dx.doi.org/10.1016/j.tra.2015.05.005
8. D'Ovidio, F.D., Leogrande, D., Mancarella, Schinzano, A. and Viola, D. (2014) A Multivariate Analysis of the Quality of Public Transport Services. Procedia—Economics and Finance, 17, 238-247. http://dx.doi.org/10.1016/S2212-5671(14)00868-5
9. Wang, Y., Guo, J., Ceder, A., Currie, G., Dong, W. and Yuan, H. (2014) Waiting for Public Transport Services: Queueing Analysis with Balking and Reneging Behaviours of Impatient Passengers. Transportation Research Part B: Methodological, 63, 53-76. http://dx.doi.org/10.1016/j.trb.2014.02.004
10. Nicolás, D.S., Colomer, O., Fiore, M. and Colomer, J.V. (2014) Restructuring of Public Transport Service of Alcoy. Procedia—Social and Behavioral Sciences, 160, 410-419. http://dx.doi.org/10.1016/j.

sbspro.2014.12.153

11. Bruglieri, M., Colorni, A., Lu□, A., Bruschi, F., Rana, V. and Nocerino, R. (2015) A RealTime Information System for Public Transport in Case of Delays and Service Disruptions. Transportation Research Procedia, 10, 493-502. http://dx.doi.org/10.1016/j.trpro.2015.09.003
12. Olsson, L.E., Friman, M., Pareigis, J. and Edvardsson, B. (2012) Measuring Service Experience: Applying The Satisfaction with Travel Scale in Public Transport. Journal of Retailing and Consumer Services, 19, 413-418. http://dx.doi.org/10.1016/j.jretconser.2012.04.002
13. Khalid, U.A., Bachok, S., Osman, M.M. and Ibrahim, M. (2014) User Perceptions of Rail Public Transport Services in Kuala Lumpur, Malaysia: KTM Komuter. Procedia—Social and Behavioral Sciences, 153, 566-573. http://dx.doi.org/10.1016/j.sbspro.2014.10.089
14. Song, J., Eom, J.K., Lee, K.S., Min, J.H. and Yang, K.Y. (2015) Public Transportation Service Evaluations Utilizing Seoul Transportation Card Data. Procedia—Computer Science, 52, 178-185. http://dx.doi.org/10.1016/j.procs.2015.05.053
15. Isabello, A., Pensa, S., Arnone, M. and Rosa, A. (2014) Reviewing Efficiency and Effectiveness of Interurban Public Transport Services: A Practical Experience. Transportation Research Procedia, 1, 243-252. http://dx.doi.org/10.1016/j.trpro.2014.07.024
16. Redman, L., Friman, M., Gãrling, T. and Hartig, T. (2013) Quality Attributes of Public Transport that Attract Car Users: A Research Review. Transport Policy, 25, 119-127. http://dx.doi.org/10.1016/j.tranpol.2012.11.005
17. dell'Olio, L., Ibeas, A. and Cecin, P. (2011) The Quality of Service Desired by Public Transport Users. Transport Policy, 18, 217-227. http://dx.doi.org/10.1016/j.tranpol.2010.08.005
18. Daraio, C., Diana, M., Di Costa, F., Leporelli, C., Matteucci, G. and Nastasi, A. (2015) Efficiency and Effectiveness in the Urban Public Transport Sector: A Critical Review with Directions for Future Research. European Journal of Operational Research, 248, 1-20. http://dx.doi.org/10.1016/j.ejor.2015.05.059
19. Yaliniz, P., Bilgic, S., Vitosoglu, Y. and Turan, C. (2011) Evaluation of Urban Public Transportation Efficiency in Kutahya, Turkey. Procedia—Social and Behavioral Sciences, 20, 885-895. http://dx.doi.org/10.1016/j.sbspro.2011.08.097

20. Anna, C., Antonello, D. and Angelo, P. (2014) A Panel Data Approach to Evaluate the Passenger Satisfaction of a Public Transport Service. Procedia Economics and Finance, 17, 231- 237. http://dx.doi.org/10.1016/S2212-5671(14)00708-4
21. Lai, W.-T. and Chen, C.-F. (2011) Behavioral Intentions of Public Transit Passengers—The Roles of Service Quality, Perceived Value, Satisfaction and Involvement. Transport Policy, 18, 318-325. http://dx.doi.org/10.1016/j.tranpol.2010.09.003
22. de Oña, J., de Oña, R., Eboli, L. and Mazzulla, G. (2013) Perceived Service Quality in Bus Transit Service: A Structural Equation Approach. Transport Policy, 29, 219-226. http://dx.doi.org/10.1016/j.tranpol.2013.07.001
23. Del Castillo, J.M. and Benitez, F.G. (2012) A Methodology for Modelling and Identifying Users Satisfaction Issues in Public Transport Systems Based on Users Surveys. Procedia— Social and Behavioral Sciences, 54, 1104-1114. http://dx.doi.org/10.1016/j.sbspro.2012.09.825
24. Garrido, C., de Oña, R. and de Oña, J. (2014) Neural Networks for Analyzing Service Quality in Public Transportation. Expert Systems with Applications, 41, 6830-6838. http://dx.doi.org/10.1016/j.eswa.2014.04.045
25. Massami, E.P. and Myamba, B.M. (2014) Evaluation of Road Traffic Accident Risk Based on Fuzzy Set Theory. International Journal of Emerging Technology and Advanced Engineering, 4, 724-731.
26. Gong, B., Chen, X. and Hu, C. (2012) Fuzzy Entropy Clustering Approach to Evaluate the Reliability of Emergency Logistics System. Energy Procedia, 16, 278-283. http://dx.doi.org/10.1016/j.egypro.2012.01.046
27. Bray, S., Caggiani, L., Dell'Orco, M. and Ottomanelli, M. (2014) Features Selection Based on Fuzzy Entropy for Data Envelopment Analysis Applied to Transport Systems. Transportation Research Procedia, 3, 602-610. http://dx.doi.org/10.1016/j.trpro.2014.10.039
28. Lin, R. and Tan, J. (2013) Evaluation of Port Development Based on the Theory of Driving Force and the Law of Entropy Weight. Procedia—Social and Behavioral Sciences, 96, 1774- 1783. http://dx.doi.org/10.1016/j.sbspro.2013.08.202
29. Behzadian, M., Otaghsara, S.K., Yazdani, M. and Ignatius, J. (2012) A State-of the-Art Survey of TOPSIS Applications. Expert Systems

with Applications, 39, 13051-13069. http://dx.doi.org/10.1016/j.eswa.2012.05.056

30. Awasthi, A., Chauhan, S.S., Omrani, H. and Panahi, A. (2011) A Hybrid Approach Based on SERVQUAL and Fuzzy TOPSIS for Evaluating Transportation Service Quality. Computers & Industrial Engineering, 61, 637-646. http://dx.doi.org/10.1016/j.cie.2011.04.019
31. Torlak, G., Sanal, M., Zain, S. and Sevkli, M. (2011) Analyzing Business Competition by Using Fuzzy TOPSIS Method: An Example of Turkish Domestic Airline Industry. Expert Systems with Applications, 38, 3396-3406. http://dx.doi.org/10.1016/j.eswa.2010.08.125
32. Awasthi, A., Chauhan, S.S. and Omrani, H. (2011) Application of Fuzzy TOPSIS in Evaluating Sustainable Transportation Systems. Expert Systems with Applications, 38, 12270- 12280. http://dx.doi.org/10.1016/j.eswa.2011.04.005
33. Toosi, N.M. and Kohanali, R.A. (2011) The Study of Airline Service Quality in the Qeshm Free Zone by Fuzzy Logic. The Journal of Mathematics and Computer Science, 2, 171-183.
34. Lin, Z.Z. and Wen, F.S. (2009) Entropy Weight Based Decision-Making Theory and Its Application to Black-Start Decision-Making. Proceedings of the CSU-EPSA, 21, 26-33.
35. Jing, L.P. and Michael, K.N. (2009) An Entropy Weighting K-Means Algorithm for Subspace Clustering of High-Dimensional Sparse Data. IEEE Transactions on Knowledge and Data Engineering, 19, 1026-1041. http://dx.doi.org/10.1109/TKDE.2007.1048
36. Hwang, C. and Yoon, K. (1981) Multiple Attribute Decision Making: Methods and Application. Springer, New York. http://dx.doi.org/10.1007/978-3-642-48318-9
37. Abdullah, L. and Zamri, N. (2010) Ranking of the Factors Associated with Road Accidents Using Correlation Analysis and Fuzzy TOPSIS. Australian Journal of Basic and Applied Sciences, 4, 35-48.
38. Chen, Y. and Whalley, A. (2012) Green Infrastructure: The Effects of Urban Rail Transit on Air Quality. American Economic Journal: Economic Policy, 4, 58-97. http://dx.doi.org/10.1257/pol.4.1.58

CHAPTER 17

Intermodal Terminal Localisation Using a Linear Programming Approach: The Case Study of Togo and West African Landlocked Countries

Bomboma Kalgora

School of Economics and Management, Shanghai Maritime University, Shanghai, China

ABSTRACT

In this paper, four potential cities to host an intermodal terminal for containers flowing through the Togolese transport corridor are examined. The transport cost minimization through the corridor is the main objective. Consequently, the transport modes that offer the least cost to the transport supply chain are proposed. To attain this goal the paper aims to identify the optimal location for an intermodal terminal on the Togolese corridor, by using the mathematical linear programming model. For this, three transport

Citation: Kalgora, B. (2019), "Intermodal Terminal Localisation Using a Linear Programming Approach: The Case Study of Togo and West African Landlocked Countries". *Journal of Transportation Technologies*, **9**, 215-231. doi: 10.4236/jtts.2019.92014.

scenarios are analyzed, the rail, the road, and the combination of these two transport modes to each of the landlocked countries (LLCs) capital cities of Burkina Faso, Mali and Niger. Data of the average transport cost per mode, the cargo demand of the LLCs, and the distances from origin to destinations are input in the LINGO software. Based on the optimization results, we find that among the preselected terminals, the city of Mango located at 550 km in the northern part of the country is the optimal host location for an intermodal terminal along the Togolese corridor. The results of this study may be helpful to transport policy makers in the quest of rendering better servicing to the landlocked countries.

Keywords: Transport, Optimization, Intermodal Transport, Terminal, Landlocked Country, Mathematical Linear Programming, Lingo

INTRODUCTION

In today's ever increasing demand-driven world, transport supply chain has become one of the crucial focus areas in many businesses. It is easy to underestimate the complexity of global supply chain. Yet the growth of global market, the increasing customer expectations, the raising costs, the more intense and diverse competitive pressures are driving the development of new transport strategies to cope with the intricate demand of cargo transport. Over the last 20 years or so, the growth of intra-continental trade and the observed increase of road transport inefficiencies have led to the need of examining the issue of competition with alternative modes (Snaddon D. R., 2001) [1]. Transport policies at different levels advocate rail as being a more sustainable traffic mode and therefore propose a shift of volumes from road to more energy efficient traffic modes. In this regard, Parola F., & Sciomachen A. (2005) [2] imply that the only strategic decision would be the implementation of rail for connecting seaports with hinterland through inland intermodal terminals. Those terminals are of major importance for the efficiency of the intermodal transport as well as for efficient access from/to seaports. Seaports and inland intermodal terminals are connected with high capacity traffic modes, such as rail, rather than only with road. According to Jean-Paul R. (2017) [3], the selection of a transportation mode is the outcome of a number of factors. The author mentioned factors among which the cost being an important element to take into account, but also service, frequency and the general value of time attributed to the goods being transported. The attractiveness of a specific mode in relation to others is hence a trade-off between cost and value of time.

Though the causality between transport and wealth generation is not specific in the literature, transportation developments that have taken place since the beginning of the industrial revolution have been linked to growing economic opportunities. The role of infrastructure and reliable efficient transport in ensuring the movement of people and goods is obviously central for the development of countries. At each stage of societal development, a particular transport technology has been developed or adapted with an array of impacts. There has been significant growth in research on modal transport in freight distribution since the 1990s (Anny-del-Mar A.-A. & José M.-F., 2017) [4]. Improving transportation and especially intermodal transport has become a major goal in modern transport policy. In order to enhance logistics efficiencies and boost international trade in Africa, since the mid-1990s governments have embarked on formulating strategic reforms, particularly in the ports and inland transportation, in collaboration with lending organizations like the World Bank and African Development Bank (UNCTAD, 2003) [5].

The transport logistics accounts for as much as 30% of the total cost of logistics operations, and is by far the largest component of the cost structure of a business' logistics (Chang Y.H., 1998) [6]. The inland transportation, happens to cost more in developing countries than in developed ones. These countries show a significant cost gap when compared to OECD countries. A poor transport service level can negatively affect the competitiveness of regions and corporations and thus have a negative impact on the regional added value and employment. The lack of transport infrastructures can be seen as a constraining factor on development. Although transport by itself is not a sufficient condition for development, in developing economics, the lack of transportation infrastructures and regulatory impediments are jointly impacting economic development by not only conferring higher transport costs, but also delaying rendering supply chain management to make them unreliable (Jean-Paul R., 2017) [3]. For these reasons, the overall cost of transportation and, by extension, doing business in African countries is much higher than in other developing regions like Latin America and Asia (Amjadi A., & Yates A., 1995) [7].In fact, in view of the crucial role transport supply chain plays, the need to reduce inland logistics costs is more acute in developing countries. To this end, Limão N., & Venables A. J. (2001) [8] conclude that the high costs of transportation in Sub-Sahara Africa accounts for much of the region's poor economic performance. They report that a landlocked country's transport costs can be up to 55 percent higher than those of a comparable coastal country.

To cope with the ever growing demand for cargo transport and consequently the increasing containership size (Kalgora B. & Christian T.M., 2016) [9], many developed countries have attained substantial transport costs cut on their inland haulage by implementation of optimized transportation in the transport supply chain. This transport optimization mainly consisted on the introduction of transport intermodality, thus attaining substantial economies of scale in the inland transport main haulage. In line with this evolution, and this ever increasing trade volume, developing regions in the world such as the West Africa transport system have been of recent topic on how to minimize the inland transport cost henceforth from coastal to landlocked countries.

However, a research investigating the intermodal transport and the intermodal terminal localization in Sub-Saharan Africa, especially for West Africa countries which have seaports and are servicing the landlocked countries of Burkina Faso, Mali and Niger has not been found in the literature. Therefore the study attempt to determine the optimal intermodal terminal for cargo transport flowing through the Togolese corridor to and from the hinterland mentioned countries, using the mathematical linear programming function computed in the LINGO software. The remainder of this study is organized in following way. Section 2 discusses the literature review pertaining of the intermodal terminal and intermodal transport. Section 3 presents the data and the network modelling of the intermodal transport. Section 4 elaborates the mathematical objective function of the intermodal terminal localization, while Section 5 presents the obtained results. Then the study concludes in Section 6 with a suggestion approach toward policy makers.

LITERATURE REVIEW

Intermodal Terminal and Transport Definition in the Literature

Several authors came up with various definition of the intermodal terminal. As the interface point between rail and road, they are critical elements in the total freight distribution chain (Luis F., & Julius S., 1995) [10]. Bart W.W., et al., (2007) [11] and the UNCTAD (2004) [12], also identified terminals as nodes, interchange or articulation points, where different transport links meet. They serve as consolidation nodes in intermodal transportation network where containers are transferred between freight trains and trucks (Peng

G., et al., 2018) [13]. The role of intermodal terminal has also increased significantly through the implementation of intermodal technologies and, over time, terminals have become increasingly technically and technologically complex systems. Consequently, the European Conference of Ministers of Transport (1993) [14] and Macharis C, & Bontekoning Y.M. (2004) [15] define intermodal freight transport as the combination of at least two modes of transport in a single transport chain, without a change of container for the goods, with most of the route travelled by rail, inland waterway or ocean-going vessel and with the shortest possible initial and final journeys by road. Transportation costs constitute a significant portion of total transport supply chain spending (Fredrik Bärthel & Johan, 2004) [16], henceforth intermodal freight transport works best within markets with relatively large flows occurring over long distances with the aim of reducing the costs. It depends on four major factors: location, efficiency, financial sustainability and rail level of service (Meyrick & Associates, 2006) [17], of which the terminal localization is one of the most crucial factors.

The Intermodal Terminal Localization Approach in the Literature

Most terminal location research has focused primarily on problems in which each demand point is uniquely allocated to a single terminal, meaning that, each demand point can send and receive via only a single terminal (Limbourg S., & Jourquin B., 2008) [18]. A large part of models dealt with the problem of designing intermodal terminals, including the planning for their location. Even with regard to the problem of optimizing the location, there are many different models using different methods (Mayer G., & Wagner B., 2002) [19]. Early researchers on the topic have addressed continuous (planar) and discrete terminal location problems from a location theory perspective. Likewise, O'Kelly M.E. (1986) [20] presented a Weiszfeld type solution procedure for continuous terminal location with one and two terminals. Others continuous hub location research have considered a single hub minimax problem (O'Kelly M.E., & Harvey J.M., 1991) [21], planar and spherical problems (Aykin T., & Brown G.F., 1992) [22] and problems with squared Euclidean distances (O'Kelly M.E., 1992) [23]. Moreover, Klincewicz J.G. (1991) [24] has considered several exchange heuristics, and has shown promising results using Tabu search and GRAP (greedy randomized adaptive search procedure) heuristics for single allocation intermodal location problems (Klincewicz J.G., 1992) [25]. Skorin-Kapov D. & Skorin-Kapov J. (1994) [26] analyze a more comprehensive Tabu search heuristic and O'Kelly, M.E. et al. (1995) [27] developed lower bounds for hub location problems. Weber's theory of location is an example of lowest costs optimization based on minimizing transport cost.

Koen H. van Dam, et al. (2007) [28], in their paper develop an integral model for the evaluation of road-rail intermodal freight hub location decisions. The authors' research focuses on road-rail intermodal transport which is contestable with road only options over longer distances. Their research results showed that the transport by train is cost-effective for transportation over a long distance and with large quantities while road is more flexible in its route and accessibility for collection and distribution activities for medium to short distances. Also Pekin E., & Macharis C. (2007) [29] used a GIS based location analysis model for intermodal terminals, while Kayikci Y. (2010) [30] applies a fuzzy AHP and artificial neural networks methods in the process of decision-making to select the most appropriate location for an intermodal freight centre. Bogusz W., et al. (2016) [31], proposed the location of intermodal terminals including port, regional, local and transit terminals. Their work also gave the forecast of intermodal units' transshipments by the year 2030. They also proposed organizational solutions that provide for significant improvement to the quality of intermodal transport. In the case of rail-road transport terminals, their work suggested that, services should cover up to 150 km using the road transport means. From a practical point of view, one should select a location in such place that the area of operation of all the terminal covers the largest possible area of the country.

Other research in the field used linear programming for a set of estimated potential locations as input for an iterative procedure based on the p-hub median problem (Limbourg S., & Jourquin B., 2008) [18]. Their paper focused on hub location in systems involving transportation of passengers or goods. They concluded that the performance of the optimal configuration that results from their model (for the same number of hubs) is more than three times better in terms of reduction of the ton · km/year transported by road. This solution would achieve 35% of the Marco Polo I annual programme objective which aims at achieving a traffic shift or avoidance. This is a substantial part of the expected yearly aggregate increase in international road freight traffic, measured in tonne-kilometres, to modes of transport with lower external costs in which road journeys are as short as possible Marco Polo II (2013) [32]. A shift from the road to more environmentally friendly modes, introduced by the European Commission (2001) White Paper [33]. Arnold P., et al. (2004) [34] used simulated annealing to find near-optimal locations of terminals. The resulting linear integer program was solved heuristically. The model was illustrated for the location of rail/road terminals in the Iberian Peninsula. In this application, the impact of variations in the supply of transport on modal shares of containerized freight transport was explored.

Generally, throughout the literature pertaining to this field of research, few studies used the linear programming for the intermodal terminal localization. Furthermore, there is no trace of previous pieces of literature on the topic of intermodal terminal localization for Togo, to a larger sense in West Africa.

The Methodological Approach of an Intermodal Terminal Localization

The process begins with an analysis of potential location sites. A new freight hub has to be economically viable so a certain transport volume has to be reached with a good price for transport and handling to ensure the new intermodal freight terminal is a beneficial investment. The development of the hinterland intermodal terminal requires innovative solutions so as to shift the freight to the railway, good railway services and the introduction of block trains (Jencek P., & Twrdy E., 2008) [35]. For the development of intermodal transport a sufficiently large concentration of freight is needed in one place.

The platforms should be set up and implemented in accordance with the regional demands and taking into consideration all of the important influential factors and their mutual interactions. For this, the following need to be performed:

- calculating the potential terminal throughput,
- defining the optimal type of platform,
- defining the optimal location,
- integrating the existent non-connected platforms into the functional network,
- calculating the transportation effect of the platforms,
- analysing the transportation effects,
- determining environmental effects,
- determining the possibilities for operation and further development of platforms.

The analysis of the regional needs for a freight platform is of utmost importance and represents the basis for reaching all subsequent decisions. The volume of freight handled that can be realistically expected at an individual terminal is very important for the development of the regional terminal. This depends on the function developed by the terminal, the terminal catchment area and of course on the attractiveness of the individual location. Jencek P., & Twrdy E. (2008) [35] gave recommendations concerning the volume—TEU (Twenty-foot equivalent unit)—throughput (see Table 1). They suggest in terms of tonnage, that the minimum freight flows in the terminal catchment area should be 4 million tons/year. This would enable the formation of 1 block train daily and would justify the development of an intermodal terminal.

DATA AND MODELLING

Data

Table 2 shows the import and export freight demand from 2005 to 2016 summed up for each destination. The data was obtained from the Burkinabe, Mali and Niger's Chamber of Commerce and PMAWCA (2016) [36].

The research forecast the cargo volume flow for the next 10 years (from 2017 to 2026). To attain this forecast, the researcher proceed by simply calculating the rate of evolution by variation period of n and then the overall average evolution rate which are both mathematically presented as followed:

Calculation of Rate Evolution by variation of year,

$$T_m = \frac{t_{G1} + t_{G2} + \cdots + t_{Gn}}{n} \qquad (1)$$

t_G : Rate of evolution by variation period of (n).

V_i : Initial volume of year (n).

V_j : Final volume of year (n + 1).

Calculation of Overall Average Evolution Rate,

$$T_m = \frac{t_{G1} + t_{G2} + \cdots + t_{Gn}}{n} \qquad (2)$$

T_m : Overall Average Evolution Rate.

t_{G1} Rate of evolution by variation period from (n) to (n + 1).

t_{Gn} Rate of evolution by variation period over (n) period.

n: number of variation period.

Evolution Rate of demand obtained is on average 10.9 percent on yearly basis. This result meets our expectations as the world organization of road transport (IRU)[1] mentioned it in their report of fact sheet typical problems of intermodal terminals, that some terminals have an average growth rate of up to 10% on yearly basis. From these mathematical calculations, we obtain the results presented in Table 3 of which the forecasted volume starts from the year 2017. Because of the spatial constraints and the challenges of extension of the terminal in the future, the storage space may become a limiting factor. Therefore this study is using the forecasted demand of the year 2023 as a benchmark. The year 2023 is selected as the benchmark year based on the Togolese PND (Plan National de Developpement) for the period 2018-2022. The PND is a national development plan which aims at the structural transformation of the Togolese economy. Its first three effects of the expected impacts in ax 1 of the plan document focus on the transport sector (Nadoun C., 2019) [37]. The forecasted imports and exports demand is used to determine the optimal intermodal terminal location along the Lomé-Ouagadougou, Niamey and Bamako transport corridors routes.

Table 1. Intermodal terminal throughput limit recommendation.

Terminal Size	Terminal Throughput Limit
Small terminals	5000 TEU/year
Medium terminals	5000 to 20,000 TEU/year
Large terminals	20,000 to 40,000 TEU/year
Super large terminals	more than 40,000 TEU/year

Source: Adapted from Jencek P., & Twrdy E. (2008).

[1]Union Internationale des Transports Routiers, https://www.iru.org.

Table 2. Demand imports & exports data for selected LLCs in W/A (million tons).

LLC	Year										
	2006	2007	2008	2009	2010	2011	2012	2013	2014	2015	2016
Burkina Faso (Ouaga-dougou)	1.87	2.26	2.6	2.28	3.31	4.76	5.52	5.99	5.77	6.33	7.02

Mali (Bamako)	3.9	4.54	4.82	3.78	4.76	5.11	5.89	5.99	6.06	6.46	7.16
Niger (Niamey)	1.58	1.78	2.26	2.79	3.11	3.46	3.34	3.6	3.94	5.21	5.79
Total	7.35	8.58	9.68	8.85	11.18	13.33	14.75	15.58	15.77	18	19.97

Source: Burkinabe, Mali and Niger's Chamber of Commerce and PMAWCA.

Table 3. Forecasted demand imports & exports data for selected LLCs in W/A (millions of tons).

LLC	Year									
	2017	2018	2019	2020	2021	2022	2023	2024	2025	2026
Burkina Faso (Ouagadougou)	7.79	8.64	9.58	10.62	11.78	13.07	14.49	16.07	17.83	19.77
Mali (Bamako)	7.94	8.81	9.77	10.83	12.02	13.33	14.78	16.39	18.18	20.16
Niger (Niamey)	6.42	7.12	7.90	8.76	9.72	10.78	11.95	13.26	14.70	16.31
Total	22.15	24.56	27.24	30.22	33.51	37.17	41.22	45.72	50.71	56.24

Source: Processed by the Author.

Consequently, the proposed optimization model takes into account five types of parameters: the spatial parameter compiled in Table 3 (cargo's demand from the LLCs), the transport mode (rail and road), the preselected intermodal terminals and the distance between nodes (see Appendix 1 & 2). The average transport cost is in USD per ton per kilometre. This average transport price mechanism is calculated based on the fuel cost, road conditions, as well as any indirect and direct costs associated with inland transportation (Borderless, 2016) [38].

The upper limit on the freight rates is fixed by the road freight rates; i.e. the railway cannot set the freight rates higher than the road freight rates unless customers are willing to pay more than the road freight rates. Rajiv D., et al. (2011) [39] suggest that the freight rate should be set in sort to determine the optimum pricing policy and terminal location in order to guarantee a certain level of profit for the railway from the intermodal service; i.e. pricing should be set in such a way that the railway can achieve this profit without losing traffic to road transport. Henceforth this study applies the freight rate of 0.14 USD/ton/km as same as the one from Ivory Coast, added to the fact that the two countries share numerous similarities.

Intermodal Transport Modeling

It is then necessary to determine the location of the terminal to open as well as the path of the cargo streams so as to minimize the total transport cost.

We designate by:

OD: all the cargo from origin to destination,

O: the origin (international gateway),

D: all destinations (3 LLCs),

T: all the preselected terminals.

Figure 1 illustrates the representation of the intermodal transport. The mathematical optimization model is formulated as an integer linear programming problem with the minimization of transport costs as objective function (Schrijver A., 1998) [40].

INTERMODAL TERMINAL LOCATION OBJECTIVE FUNCTION

The decision to open a terminal or not can be represented by a binary variable y_i which will be worth 1 if the terminal must be open and 0 otherwise. Similarly, the cargo stream X from an origin o to a terminal i, by type of transport mode r, on the one hand, and a terminal i to a destination j and by type of transport mode r will be represented by variables of streams X^p_{oir} and X^p_{ijr} . The average transport cost C of p from the origin o to the terminal i and by type of transport mode r, on the one hand and secondly from a terminal i to a destination j per type of transport mode r is represented by transport cost variables C^p_{oir} and C^p_{ijr} On one hand the distance (d) separates the transport point of (p) from the origin (o) to a terminal (i) and by type of transport mode (r), and yet on the other hand, the transport distance of (p) from a terminal (i) to a destination (j) and by type of transport mode (r) is appropriate. It is represented by dpoirdoirp and dpijrdijrp . The model derived from all these notations is the following:

$$\min = \sum_o \sum_i \sum_r C^p_{oir} d^p_{oir} X^p_{oir} + \sum_i \sum_j \sum_r C^p_{ijr} d^p_{ijr} X^p_{ijr} \quad (3)$$

Constraints:

$$X_{oir}^{p} \leq y_i M$$

$$X_{ijr}^{p} \leq y_i M$$

$$\sum_{i} y_i = 1$$

$$\sum_{i} \sum_{f} X_{ijr}^{p} = a_i$$

$$X_{oir}^{p} \geq 0; X_{ijr}^{p} \geq 0$$

$$y_i = \{0,1\}$$

Set:

r = {1, 2} Transport modes

i = {1, 2, 3, 4} Preselected terminals

j = {1, 2, 3} Destinations

M Express infinity

a_i Express destination cargo volume demand

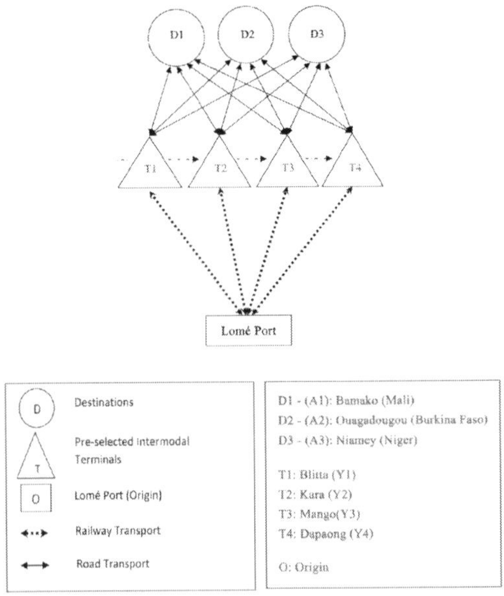

Figure 1. Intermodal transport layout. Source: Processed by the Author.

The objective function of the model contains the average cost of transportation USD$/ton/km and by mode of transport, the origin distance hub and mode of transport passing through an intermodal terminal and the volume of cargo transported from origin to destination all also passing through an intermodal terminal. The model was implemented in "LINGO" by LINDO Systems[2]. Designed to model and solve linear, nonlinear, quadratic, integer and stochastic optimization problems. Table 4 presents the basic elements of mathematical modelling language syntax of "LINGO".

Optimization occurs after the implementation of the model in the LINGO mathematical modelling language (Figure 2). Optimization results are shown in Table 5. The optimization process involves finding the global solution for the specific data, which in this case means the best location for the intermodal terminal in terms of lowering transport costs to satisfy LLCs transport needs through the corridor.

[2]https://www.lindo.com.

THE INTERMODAL TERMINAL LOCALIZATION RESULTS

The optimization results are shown in Table 5. The most optimal location for the intermodal terminal is Y_3 which represents the city of Mango is located at 550 km away from the Lomé Port, in the northern part of Togo.

Table 4. The basic syntax of "LINGO" mathematical modelling language.

Mathematical nomenclature	LINGO syntax
Minimum	MIN =
$\sum Z_{ojr}$	@sum(ORDER(o,j,r))
j = 1. M for each destination (j) in the set of destinations	@for(DESTINATION (j))
•	*
=	=
$Y \in \{0,1\}$	@bin(y)

Table 5. Terminal location optimized results.

Global optimal solution found.	
Objective value:	0.2149185E+11
Objective bound:	0.2149185E+11
Infeasibilities:	0.1862645E-08

Extended solver steps:	2		
Total solver iterations:	12		
	Variable	Value	Reduced Cost
	M	9999999.	0.000000
	Y(1)	0.000000	0.000000
	Y(2)	0.000000	−0.8839999E+10
	Y(3)	1.000000	0.000000
	Y(4)	0.000000	−0.7139999E+10
	A(1)	0.1449104E+08	0.000000
	A(2)	0.1478004E+08	0.000000
	A(3)	0.1195201E+08	0.000000
	X1(1, 1, 1)	0.000000	263.0000
	X1(1, 1, 2)	0.000000	263.0000
	X1(1, 2, 1)	0.000000	411.0000
	X1(1, 2, 2)	0.000000	411.0000
	X1(1, 3, 1)	0.000000	550.0000
	X1(1, 3, 2)	0.000000	550.0000
	X1(1, 4, 1)	0.000000	620.0000
	X1(1, 4, 2)	0.000000	620.0000
	X2(1, 1, 1)	0.000000	1332.000
	X2(1, 1, 2)	0.000000	1332.000
	X2(1, 2, 1)	0.000000	1254.000
	X2(1, 2, 2)	0.000000	1254.000
	X2(1, 3, 1)	0.000000	464.0000
	X2(1, 3, 2)	0.000000	464.0000
	X2(2, 1, 1)	0.000000	975.0000
	X2(2, 1, 2)	0.000000	975.0000
	X2(2, 2, 1)	0.000000	357.0000
	X2(2, 2, 2)	0.000000	357.0000
	X2(2, 3, 1)	0.000000	0.000000
	X2(2, 3, 2)	0.000000	0.000000

	X2(3, 1, 1)	9999999.	0.000000
	X2(3, 1, 2)	4491043.	0.000000
	X2(3, 2, 1)	9999999.	0.000000
	X2(3, 2, 2)	4780038.	0.000000
	X2(3, 3, 1)	9999999.	0.000000
	X2(3, 3, 2)	1952014.	0.000000
	X2(4, 1, 1)	0.000000	433.0000
	X2(4, 1, 2)	0.000000	433.0000
	X2(4, 2, 1)	0.000000	364.0000
	X2(4, 2, 2)	0.000000	364.0000
	X2(4, 3, 1)	0.000000	0.000000
	X2(4, 3, 2)	0.000000	0.000000

Source: LINGO Solver Results.

```
                         sets:
hub/1..4/:y;
origin/1/;
destination/1..3/:a;
  mode/1,2/;
links1(origin,hub,mode):x1;
links2(hub,destination,mode):x2;
links3(origin,hub):d1;
links4(hub,destination):d2;
                        endsets
min=
@sum(origin(o):@sum(hub(i):@sum(mode(r):d1(o,i)* x1(o,i,r))))+
@sum(hub(i):@sum(destination(j):@sum(mode(r):d2(i,j)*x2(i,j,r))));

@for(origin(o):@for(hub(i):@for(mode(r):x1(o,i,r)<=M*y(i))));
@for(hub(i):@for(destination(j):@for(mode(r):x2(i,j,r)<=M*y(i))));
@sum(hub(i):y(i))=1;
@for(destination(j):@sum(mode(r):@sum(hub(i):x2(i,j,r)))=a(j));
@for(hub(i):@for(destination(j):@for(mode(r):x2(i,j,r)>=0)));
@for(origin(o):@for(hub(i):@for(mode(r):x1(o,i,r)>=0)));
@for(hub(i):y(i)>=0);
@for(hub(i):@bin(y(i)));

data:
M=9999999;
d1=263
411
550
620;
d2=1710 1562 1423 1353
   665 517 378 308
   959 811 672 602;
a=14491041.6441944
14780036.7766997
11952012.9800406;
enddata
```

Figure 2. Part of the file scm.lng (Transport supply chain cost optimization model in LINGO). Source: Processed by the Author.

Figure 3, is a geographical illustration of the intermodal terminal location on the Togolese corridor with the city of Mango encircled in red.

Based on computational experiments, results, analysis, the research can state that the proposed model and its implementation ensure a very large range of application. First, to determine the best location for intermodal terminal along the corridor which offers an adequate solution for decision support in transport supply chain management. Second, minimize the inland total transport cost through the corridor, thus satisfying the needs of customers, and this based on the determination of the optimal location for the intermodal terminal.

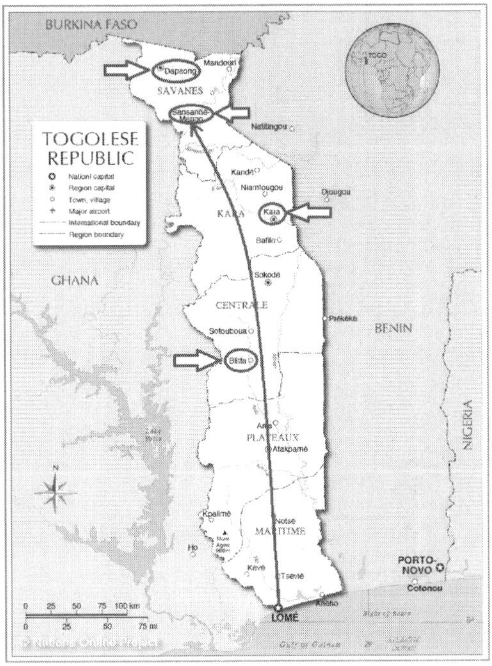

Figure 3. Map of Togo with optimal intermodal terminal. Source: Processed by the Author.

CONCLUSIONS

The choice of location for an intermodal transport terminal is an important component in a regional transport supply chain system and a paramount decision for investors. It is of crucial importance in the logistic industry and research on the topic revealed that it deserves particular attention from practitioners. The reason lies in the fact that assessing the construction resources, time, and cost involved. If decision is made to build a logistic intermodal terminal at a specific location, but there is something faulty in the decision-making process, henceforth there is no question of destroying it within a few days after finding.

The paper developed the intermodal terminal network framework for the case study of the intermodal transport through the Togolese corridor. The present framework consisted of determining the optimal intermodal terminal location using the mathematical linear programming model computed in the LINGO software. This involved the modelling of the transport network, while taking into account the demand forecast of year 2023 of the 3 LLCs. This demand forecast was required to determine the viability of such a project. This research adds value to the transport sector in Togo and is unique in the sense that it's the first of its type to identify an optimal intermodal terminal location on the Togolese corridor through the use of an optimization method of mathematical linear programming. The present research identified the city of Mango as the optimal location for the intermodal terminal. The Mango city is located 550 km north of Lomé (Capital city of Togo). The optimization results reveal the optimal intermodal terminal location in terms of total transport cost minimization for containers cargo in direction to 3 LLCs of Burkina Faso, Mali and Niger. In line with the findings and in regard to the increasing trade volume due to the ever growing demand for transport cargo to and from the landlocked countries of Burkina Faso, Mali and Niger, this case study highlights the merits of the intermodal transport for the Togolese transport industry and, advises policy makers to take into consideration its results.

CONFLICTS OF INTEREST

The author declares no conflicts of interest regarding the publication of this paper.

APPENDICES

Appendix 1. Preselected intermodal terminals and distance between nodes.

Corridor Section	Nodes	Distance (km)
Port-Preselected Terminal	Lome-Blitta (Y1)	263
	Lome-Kara (Y2)	411
	Lome-Mango (Y3)	550
	Lome-Dapaong (Y4)	620

Terminal-LLC	Blitta-Bamako	1710
	Kara-Bamako	1562
	Mango-Bamako	1423
	Dapaong-Bamako	1353
	Blitta-Ouagadougou	665
	Kara-Ouagadougou	517
	Mango-Ouagadougou	378
	Dapaong-Ouagadougou	308
	Blitta-Niamey	959
	Kara-Niamey	811
	Mango-Niamey	672
	Dapaong-Niamey	602

Source: Adapted from Google map.

Appendix 2. Average transport cost per mode.

Country	**Transport cost-road (USD/ton/km)**	**Transport cost-rail (USD/ton/km)**
Cote d'Ivoire	0.17	0.14
Benin	0.18	n/a
Nigeria	0.2	n/a
Togo	0.16	0.14
Ghana	0.14	n/a

Source: Adapted from Borderless (2016).

REFERENCES

1. Snaddon, D.R. (2001) Competition in Transportation—A Literature Analysis. Technovation, 21, 375-383. https://doi.org/10.1016/S0166-4972(00)00053-5
2. Parola, F. and Sciomachen, A. (2005) Intermodal Container Flows in a Port System Network: Analysis of Possible Growths via Simulation Models. International Journal of Production Economics, 97, 75-88. https://doi.org/10.1016/j.ijpe.2004.06.051
3. Jean-Paul, R. (2017) The Geography of Transport Systems. Fourth Edition, Routledge, New York, 440 p.
4. Anny-del-Mar, A.-A. and José, M.-F. (2017) Intermodal Transport in Freight Distribution: A Literature Review. Journal of Transport Review, 37, 782-807.
5. UNCTAD (2003) African Ports: Reform and the Role of the Private Sector. Secretariat of UNCTAD (UNCTAD/SDTE/TLB/5).
6. Chang, Y.H. (1998) Logistical Management. Hwa-Tai Bookstore Ltd.
7. Amjadi, A. and Yates, A. (1995) Have Transport Costs Contributed to the Relative Decline of African Exports? Some Preliminary Evidence. Policy Research Working Paper No. 1559, World Bank, Washington DC.
8. Limão, N. and Venables, A.J. (2001) Infrastructure, Geographical Disadvantage, Transport Costs and Trade. The World Bank Economic Review, 15, 451-479. https://doi.org/10.1093/wber/15.3.451
9. Kalgora, B. and Christian, T.M. (2016) The Financial and Economic Crisis, Its Impacts on the Shipping Industry, Lessons to Learn: The Container-Ships Market Analysis. Open Journal of Social Sciences, 4, 38-44. https://doi.org/10.4236/jss.2016.41005
10. Luis, F. and Julius, S. (1995) Measuring the Performance of Intermodal Freight Terminals. Journal Transportation Planning and Technology, 17, 269-280. https://doi.org/10.1080/03081069308717517
11. Bart, W.W., Enno, M. and Peter, N. (2007) Intermodal Freight Terminals: An Analysis of the Terminal Market. Journal of Transportation Planning and Technology, 23, 105-128.
12. UNCTAD (2004) Assessment of a Seaport Land Interface: An Analytical Framework. UNCTAD Secretariat, UNCTAD/SDTE/TLB/MISC/2004/3.
13. Peng, G., Wenming, C., Yi, W. and Nils, B. (2018) Gantry Crane Scheduling in Intermodal Rail-Road Container Terminals. Interna-

tional Journal of Production Research, 56, 5419-5436. https://doi.org/10.1080/00207543.2018.1444812
14. European Conference of Ministers of Transport (1993) Terminology on Combined Transport. http://www.cordis.lu/transport
15. Macharis, C. and Bontekoning, Y.M. (2004) Opportunities for OR in Intermodal Freight Transport Research: A Review. European Journal Operation Research, 153, 400-416. https://doi.org/10.1016/S0377-2217(03)00161-9
16. Bärthel, F. and Johan (2004) Developing Intermodal Transport for Small Flows over Short Distances. Journal of Transportation Planning and Technology, 27, 403-424.
17. Meyrick and Associates (2006) National Intermodal Terminal Study. Australian Department of Transport and Regional Services, Publication 10781.
18. Limbourg, S. and Jourquin, B. (2008) Optimal Rail-Road Container Terminal Locations on the European Network. Transportation Research Part E, 45, 551-563. https://doi.org/10.1016/j.tre.2008.12.003
19. Mayer, G. and Wagner, B. (2002) Hublocator: An Exact Solution Method for the Multiple Allocation Hub Location Problems. Computers and Operations Research, 29, 715-739. https://doi.org/10.1016/S0305-0548(01)00080-6
20. O'Kelly, M.E. (1986) The Location of Interacting Hub Facilities. Transportation Science, 20, 92-106. https://doi.org/10.1287/trsc.20.2.92
21. O'Kelly, M.E. and Harvey, J.M. (1991) Solution Strategies for the Single Facility Minimax Hub Location Problem. The Journal of the Regional Science Association International, 70, 367-380. https://doi.org/10.1007/BF01434594
22. Aykin, T. and Brown, G.F. (1992) Interacting New Facilities and Location-Allocation Problem. Transportation Science, 3, 212-222. https://doi.org/10.1287/trsc.26.3.212
23. O'Kelly, M.E. (1992) A Clustering Approach to the Planar Hub Location Problem. Annals of Operations Research, 40, 339-353. https://doi.org/10.1007/BF02060486
24. Klincewicz, J.G. (1991) Heuristics for the p-Hub Location Problem. European Journal of Operation Research, 53, 25-37. https://doi.org/10.1016/0377-2217(91)90090-I
25. Klincewicz, J.G. (1992) Avoiding Local Optima in the p-Hub Location Problem Using Tabu Search and GRASP. Annals of Operations

Research, 40, 283-302. https://doi.org/10.1007/BF02060483
26. Skorin-Kapov, D. and Skorin-Kapov, J. (1994) On Tabu Search for the Location of Interacting Hub Facilities, Forthcoming in European. Operational Research, 73, 502-509. https://doi.org/10.1016/0377-2217(94)90245-3
27. O'Kelly, M.E., et al. (1995) Lower Bounds for the Hub Location Problem. Location Science, 41, 713-721. https://doi.org/10.1287/mnsc.41.4.713
28. van Dam, K.H., Zofia, L., Luis, F. and Ackchai, S. (2007) Planning the Location of Intermodal Freight Hubs: An Agent Based Approach. IEEE International Conference on Networking, Sensing and Control, London, 15-17 April 2007, 187-192. https://doi.org/10.1109/ICNSC.2007.372774
29. Pekin, E. and Macharis, C. (2007) A GIS-Based Location Analysis Model for Intermodal Terminals. International Logistics and Supply Chain Congress, Istanbul.
30. Kayikci, Y. (2010) A Conceptual Model for Intermodal Freight Logistics Centre Location Decisions. The Asian Journal of Shipping and Logistics, 27, 331-335. https://doi.org/10.1016/j.sbspro.2010.04.039
31. Bogusz, W., Dariusz, M., Leszek, C. and Igor, II. (2016) The Concept of the Development of Intermodal Transport Network Illustrated by Polish Market. NAŠE MORE: Znanstveni časopis za more i pomorstvo, 64, 33-37. https://doi.org/10.17818/NM/2017/1.6
32. Polo, M. II (2013) Annual Work Programme. Annex 1.
33. European Commission (2001) European Transport Policy for 2010. Time to Decide. COM 370 Final.
34. Arnold, P., Peeters, D. and Thomas, I. (2004) Modelling a Rail/Road Intermodal Transportation System. Transportation Research Part E, 40, 255. https://doi.org/10.1016/j.tre.2003.08.005
35. Jencek, P. and Twrdy, E. (2008) Development of Regional Transport Logistics Terminal. Transport Logistics Approach, 4, 239-249.
36. PMAWCA (2016). http://www.agpaoc-pmawca.org
37. Nadoun Coulibaly (2019) Togo: Faure Gnassingbé lance officiellement son PND. Economie and Finance, Politique Economique du 03 mars 2019 à 16h06, Jeune Afrique. https://www.jeuneafrique.com/743441/economie/togo-faure-gnassingbe-lance-offici elle-ment-son-plan-national-de-developpement
38. Borderless (2016) Regional Integration through the Development

of Trade and Transport. Executive Secretariat, Borderless Alliance.

39. Rajiv, D., Rabindra, N.B., Behzad, G. and Aditya, P. (2011) Optimal Pricing and Terminal Location for a Rail-Truck Intermodal Service—A Case Study. International Journal of Logistics Research and Applications, 14, 335-349. https://doi.org/10.1080/13675567.2011.643778
40. Schrijver, A. (1998) Theory of Linear and Integer Programming. John Wiley and Sons, Hoboken.

CHAPTER

18

Study on Objectives and Strategies for Energy Conservation and Emission Reduction in Transportation Industry during the 13th Five-Year Plan Period in Beijing

Ying Liu[1], Xiaoyi Li[2], Jun Yang[3,4]

[1]College of Metropolitan Transportation, Beijing University of Technology, Beijing, China

[2]Energy & Environment Center, Beijing Transport Institute, Beijing, China

[3]Beijing Jiaotong University, Beijing, China

[4]Beijing Transport Institute, Beijing, China

ABSTRACT

Facing increasing passenger and cargo transport demand and limited resource in the 13th Five-Year period, how to make a breakthrough and substantial

Citation: Liu, Y. , Li, X. and Yang, J. (2017), "Study on Objectives and Strategies for Energy Conservation and Emission Reduction in Transportation Industry during the 13th Five-Year Plan Period in Beijing". *Journal of Transportation Technologies*, **7**, 304-317. doi: 10.4236/jtts.2017.73020.

progress has become a key issue on planning and the Top-level Design. In this paper we judged and analyzed the current development and potential demand of the energy saving and emission reduction in Beijing traffic industry. Through application of energy and emission prediction model which based on the vehicle activity data, the development goals of "one drop, double control, and triple upgrade" have been put forward. In order to achieve the goal, "5 + 1" development strategies should be implemented, and we also proposed the thinking and recommendations on sustainable development of transportation.

Keywords: Transport Energy Saving and Emission Reduction, 13th **Five-Year Plan, Prediction Model, Development Goals and Strategy**

FOREWORD

At the end of 2015, Beijing resident population reached 21.705 million and vehicles was totally 5.6 million units [1] , mobile development of the megacities, the unbalanced supply and demand structure of urban traffics for population expansion became increasingly prominent and external rigid constraints of energy and environment were becoming more and more urgent.

During the 13th five-year plan period, guided by the five major principles, we'll comprehensively carry forward and implement the five in one layout of socialism undertaking with Chinese characteristics and speed up the development of four transportation sectors while the cleaning energy development strategy, response to climate change and prevention and control of vehicle pollution raise higher requirement on the transportation development. In coming five years, a key period for construction of an international first-class harmonious and livable capital of harmony and comprehensive establishment of a safe, convenient, efficient, green and economic modern comprehensive transportation system, the transformation of transport industry will face double pressures: enhancing quality and efficiency while saving energy and reducing carbon emission. Therefore, how to control the growth of energy consumption and carbon emission gross in transport industry by various means while enhancing service quality will be a serious challenge facing works of transport industry energy conservation and emission reduction.

This paper makes comprehensive analysis on the status and future trend of Beijing's transport energy conservation and emission reduction. Based on the energy consumption and pollution emission estimation model of individual vehicle activity level, the author reasonably estimates the future trend and proposes the objectives of transport energy conservation and emission reduction in Beijing in coming five years; around the objectives, the paper brings out development strategy and raises some reflections and suggestions on the works of transport energy conservation and emission reduction in megacities.

BACKGROUND OF STUDY

Scope

Currently there are three statistic study measurements for Beijing transportation energy consumption pollutants as shown in Figure 1, including transportation, warehousing and postal service, car rental, motor vehicle maintenance and social traffic. Transportation, warehousing and postal service involve aviation transport, railway transport, road transport, pipeline transport, storage and postal service, water transport and handling assistance activities. Transport industry covers the 12 sectors under the transportation commission, including public charge, suburban rail transit, passenger transport, taxi, tourism inter-provincial passenger transport, passenger transport and road freight transport, etc. According to the jurisdiction, the object of this study is mainly the transport industry with social traffics in some parts. By the end of the 13th five-year plan period or 2020, in view of the transport industry energy consumption and pollutant emissions, as well as carbon emissions, with quantitative analysis, the author makes clear objectives and then carries out the study on the development.

Development Status

Status of Energy Consumption and Emission of Transport Industry

As of the end of 2015, the gross energy consumption of transport industry is 333, 6000 tons of standard coal. During the 12th five-year plan period, the transport industry in the phase of large-scale operation and construction with total energy consumption is 37.05% higher than that in 2010 and the average annual growth rate is 6.49%. In the industry, the freight transportation, the public car, taxi, rail transportation and other sectors are key energy users, respectively accounting for 31.6%, 22.0%, 17.2% and 14.7% of the total. Petrol and diesel are still the main fuels, but as the new expansion of clean energy vehicles and the strong development of rail transportation, the proportion of use of electricity and gas has gradually increased from 17.6% to 13.0% [2] .

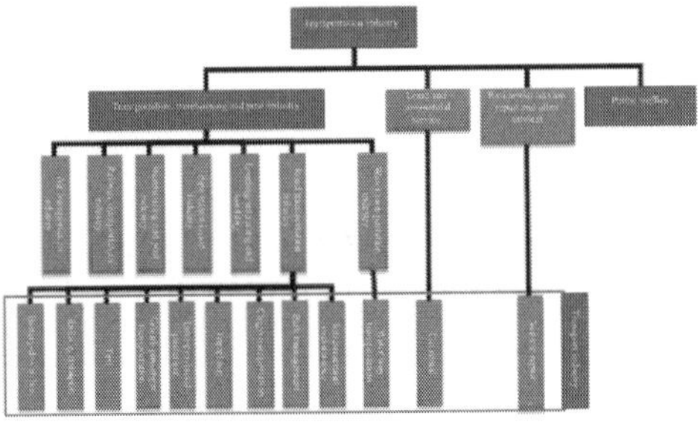

Figure 1. Structure of transport field.

According to environmental protection authority, the exhaust of vehicles is still a main source of atmospheric pollution. The data of Beijing center for transport industry energy conservation and emission reduction show, in 2014, the main contents of exhaust of vehicles are carbon monoxide and nitrogen oxides, 49.8% and 40.3% respectively. Heavy truck is an important source of PM, NOx and HC in transport industry as one of key areas of pollution emission control in the coming five years. From the gross pollution emissions in transport industry, through various measures, emissions have showed a trend of decline.

Achievements of Works during the 12th **Five-Year Plan Period**

Compared to industrial construction field with earlier development and increasingly complete system, Beijing municipal transportation's energy conservation and emission reduction work is still in its infancy. During the 12th five-year plan period by pilot demonstration, we have promoted new energy vehicles and built transportation emissions from energy consumption statistical monitoring system and made preliminary exploration to form an urban transport energy conservation and emission reduction development model with Beijing characteristics. We have made following primary achievements:

1). Green low-carbon travel gets rapid development to promote structured reduction of emissions. During the 12th five-year plan period, by encouraging public to travel with green intensive public transportation, Beijing government has optimized the structure of travel to reduce energy consumption and emissions per capita. The number of users of public transportation and passenger traffic grew from 18.86 million to 20.40 million, total length of rail transit operation increased from 336 km to 554 km, total length of bus lanes increased from 294 km to 741 km, scale of public bike rental reached 50,000 vehicles and proportion of public traffic travel in downtown rose from 39.7% to 50%. The results show that the single travel emissions per capita fell by 28% by the end of the 11th five-year plan period.

2). Transport energy structure optimization witnesses a breakthrough. At the end of 2015, the ground buses, taxi and other transport industry promoted 22,020 new energy vehicles, the growth rate of 486% in comparison with the figure at the end the 11th five-year plan period. Through new energy car index separate configuration and the non traffic limit policy, the government promoted 13,391 new energy cars among the city's private sector, reduced 340 million litres of gasoline and diesel consumption per year and 1278 tons of pollutants discharge. Industry has eliminated all yellow-labeled cars, and the proportion of vehicles with state IV standard and above standard has been increased from 38% at the end of the 11th five-year plan period to 61% at the end of the 12th five year plan period and he industry's structure of vehicle emissions has been significantly optimized.

3). Application scale of energy conservation and emission reduction product technology has been gradually expanded. Actively encourage rail transportation, transportation hubs through contract energy management mode, such as application of LED energy-saving lighting, frequency conversion energy-saving equipment like elevators, frequency conversion air conditioning, therefore during the 12th five-year plan period accumulated power saving exceeded 100 million degrees. In the aspects of construction and maintenance of the road, continue to promote energy-saving and environmentally-friendly material to make the proportion of such materials on common roads reached 85% and proportion for old pavement reached 95%. In railway energy recovery, green tires, lightweight car body and other energy saving technical reformation respects, we made breakthroughs.

4). The infrastructure capacity construction promotes the energy reduction in the transport industry. During the 12th five-year plan period Beijing launched the construction of energy consumption and emission statistics monitoring platform and transportation environment monitoring network, in combination with five-level energy consumption and emissions model system research and development, we have preliminarily implemented refinement analysis and forecast of industry' energy consumption and emissions. We have compiled and published the five industries' energy conservation standards and effectively link to Beijing municipal bureau of statistics by energy consumption statistical system innovation.

5). Exploration into transport energy conservation and emission reduction systems and mechanisms get results primarily. Beijing has established Beijing transport industry energy conservation and emission reduction work leading team and the Beijing center for transport industry energy conservation and emission reduction, according to the requirements of atmospheric pollution joint control and prevention in Beijing-Tianjin-Hebei, and the surrounding area, we have established the motor vehicle emission pollution control work coordination mechanism, actively asked the ministry of transport and Beijing for clean air and energy conservation and emission reduction of special policy and funding and support pilot demonstration projects.

6). Primarily form a social atmosphere for transport energy conservation and emission reduction. We have conducted public organizations to carry out the green travel communication campaigns, using Weibo WeChat and other new media way, we spread the concept of green traffic. For transportation enterprises we held the energy conservation and emission reduction standard stable promotion and technical training to deepen the enterprises work in energy conservation and emission reduction so as to primarily form a transport energy conservation and emission reduction atmosphere in the whole society.

We have made some achievements after past five years, but the depth and breadth of work remains to be promoted, structural reduction potential is yet to be dug out, basic ability remains to be further promoted, public consciousness in energy conservation and emission reduction needs still to be cultivated, and energy conservation and emission reduction management system will help improve adequately.

STUDY ON DEVELOPMENT OBJECTIVES

As to the study on the objective of transport energy conservation and emission reduction by the end of the 13th five-year plan period in Beijing, based on the status and future demand of Beijing in this respect, we should reasonably set development scenario, and use the energy consumption of macro calculation model quantified emissions of science to predict the future energy consumption and pollutant emissions to put forward feasible development goals.

Scenario Setting

Scenario analysis method gives full consideration to the diversity of the development in the future and takes the collective intention and will of future developing group decision-makers as one of the most important aspects of the scenario analysis [3] [4] . During the process of making the development goals, we adopt the method of scenario analysis and set the scenarios reasonably by the background including the concurrent economic development of Beijing, Tianjin and Hebei, non-capital functional evacuation and clear urban function orientation.

By the end of the 13th five-year plan period Beijing's resident population is expected to reach 23 million and the city will keep high speed growth in GDP. The active population increase and concurrent economic development of Beijing, Tianjin and Hebei cause demand of inter-city traffic to increase and commuting radius to be extended so as to enhanced the level of pluralism, humanization and refine traffic services. In the transport industry's energy consumption and pollutant forecast scenario, we consider following key factors:

1). Model structure: energy structure of the vehicles and emission structure, promotion of new energy vehicles, as well as the ascension of petrol and diesel emission standards, promotion of clean and green vehicle structure, will effectively reduce energy consumption and pollutant emissions.

2). Total car volume in the industry: according to actual operation planning new and expansion the total volume will satisfy the need for growth, it is expected the vehicle scale will be expanded by the end of the 13th five-year plan period.

3). Energy consumption (emission) strength: Ministry of Industry and Information Industry has released "Standard for the Phase IV of Fuel Consumption by Passenger Vehicles" forecasting that the fuel economics will be improved by over 20% in comparison with 2015. Beijing Environmental Protection Bureau released "Draft for Opinions for "Local Standard for the Phase VI of Fuel Consumption by Vehicles in Beijing" which will be made and released as agreed and implemented since December 1, 2017. The implementation of new emission standards can slash vehicle emissions pollution [5] . In addition, with ecological driving technology promotion, energy consumption and emissions intensity of individual vehicles will decline further.

4). Mileage: in 2020, with the increase of vehicle volume and transportation efficiency in the industry, it is predicted the average annual mileage will be equal to the figure at the end of the 12th five-year plan period basically.

Outlook of transport industry development in 2020: urban green travel ratio will reach 75%, rail transit operation mileage will reach 900 kilometers or above. The new clean energy bus proportion will be 65% or more; new vehicle rental industry will exclusively use new energy vehicles; in shipping, tourism and other industries, we'll promote new energy vehicles in combination with characteristics of industry. It is expected in 2015 the total vehicle volume in the industry will increase by more than 10% and the total mileage remain unchanged.

Designs of Modeling

At the city scale, there are three respects influencing transportation energy consumption: urban morphology (features) of land use, traffic system characteristics (traffic network accessibility, convenience and quality of service) and social economic factors (characteristics of personal and family) [6] . To accurately predict the energy consumption and pollutant emission of transport industry by the end of the 13th five-year plan period, we set up the macro listing model based on bicycle activity level and find the influencing factors of different level are embodied in three important input variables of energy consumption (pollutants) prediction model: fleet structure, range and energy intensity (emission factor); taking microscopic transportation unit as the calculating unit, to achieve the accurate measurement of the total energy consumption, we match energy intensity and driving strength need with attributes of different models as shown in the formula below.

$$T = \sum_i N_i \times (VKT_i \times C_i)$$

T—Gross energy consumption;

Ni —Number of vehicles of different model;

VKT_i—Annual average mileage of ith model;

C_i—Fuel consumption per 100 km for ith model (pollutant emission factor).

The data of vehicle database came from the database from Beijing Traffics Administration and operative vehicle database of Beijing Transportation Bureau, we make categorization according to the vehicle energy type and vehicle weight. The data such as vehicle fuel consumption structure, activity level and fuel consumption per 100 km have been obtained through the statistics and monitoring and emission factor came from Vehicle Emission Control Center of the Ministry of Environmental Protection which has been improved based on the status of Beijing.

Objectives

The model prediction shows that during the 13th five-year plan period the total volume of Beijing transport industry energy consumption and carbon emissions will continue to increase, but growth will be under control, and pollutant emissions will make transition from increase to drop. Based on the calculation result and demand the next five years development, we propose a goal "one drop, two controls and three enhancements".

Table 1. Development goals.

	Index	**2020**	**Indicator type**
1	Proportion of reduction of pollution emission in transport industry	25%	Restrictive
2	Objective of control of gross energy consumption in transport industry (10,000 tons of standard coal)	420	Restrictive
3	Objective of control of gross greenhouse gas in transport industry (10,000 tons)	1300	Restrictive
4	proportion of new cleaning vehicles in public traffic industry	65%	Anticipative
5	Cycling rate of materials of road pavement	Above 90%	Anticipative
6	Recycling rate of sewage at highway service area	30%	Anticipative

7	Completion rate of government-level monitor and statistics platform for energy conservation and emission reduction	100%	Anticipative
8	Completion rate of energy consumption management and control of key enterprises in transport industry	100%	Anticipative
9	Installation rate of energy consumption measurement instruments on the operative vehicles in transport industry	30%	Anticipative

One drop: industry pollutant emissions significantly decrease by 25%; two controls: industry energy consumption and greenhouse gas emissions are under effective control; Three enhancements: great enhancements of a new energy use ratio, resource intensive utilization level and energy conservation and emission reduction of fine management level. Among them, total restrictive indicators like pollutant, gross energy consumption and carbon emissions, new energy vehicles scale and recycling rate of sewage at highway service area will be used as reasonable guides to the green development of the industry (Table 1).

STUDY ON DEVELOPMENT STRATEGY

To realize the objectives of energy consumption and emission control in Beijing by the end of the 13th five-year plan period, development strategies have been brought out in the respects like transport system, transport tools, infrastructures and relevant management systems, policies and product technologies to guide the conduct of transport energy conservation and emission reduction in Beijing during the 13th five-year plan period.

Construction of Green Transportation System

The optimization of urban passenger transportation system and enhancement of the level of intensive and transport efficiency are fundamental measures to achieve energy conservation and emission reduction effect.

(1) Enhance the level of public transport services to promote green low-car- bon travel.

Continue to promote the construction of urban rail transportation, complete a railway transportation network with clear function levels and reasonable hierarchical structure, dig out potential of existing lines through the energy dissipation equipment modification and the optimization of organization mode and improve railway transportation energy efficiency. Comprehensively optimize and integrate bus lines on the ground, promote construction, optimization and inter-connection of bus-only lanes and build the ground quick bus commuter system. Promote environmental improvement of walking and bicycle traffics, make great to develop the public bicycle and effectively raise the proportion of green travel. By improving the proportion of green low-carbon travel by rail, bus, walking, cycling, etc., reduce energy consumption and emissions intensity.

(2) Optimize freight transportation organization to promote construction of green freight hub stations.

Actively explore into "Internet+" application, optimize the distribution network, encourage university organization patterns like night distribution and joint distribution, improve the level of transportation efficiency and greening. Expand the scale of application of technologies like green tires, research market investment and financing mode suited to the characteristics of the freight industry, further promote application of freight energy conservation and emission reduction technology. Selection conditioned newly established or reconstructed stations, promote energy-saving building materials, electric handling transmission equipment, energy-efficient low-emission facilities and other equipment, guide the construction of charging devices of new energy vehicles, encourage the construction of yard environment monitoring system and energy management system and carry out pilot green freight yards.

Use Green Transport Equipments and Tools

Continue to optimize the structure of industry operating vehicles, promote the new energy and clean energy vehicles among ground bus and taxi industry by steps to ensure the proportion of new clean buses will be above 65% during the 13th five-year plan period. Increase the demonstrative application scale of new energy vehicles in rent freight industry. Facilitate the demonstration of new clean travel buses along some commuter lines between the city and key business areas around the city and encourage enterprises to adopt new priority clean energy vehicles. According to local environmental protection standards in Beijing, promote updates in the car industry, promote the use of low -emission vehicles above Beijing VI and higher emission standard, speed up the elimination of high-emission old motor vehicles in transport industry, promote green- oriented emissions structure of operative vehicle fleet. Cooperate with planning, and development and reform and other relevant authorities in making gas and power charging facilities planning and build functional bus special gas charging network relying on the bus terminal. Perfect social public charging facilities and provide priority for electric taxi in charging service. Actively explore into diversified gas and power charging infrastructure and service patterns.

Green Transportation Infrastructure

In the process of traffic infrastructure construction, from the early -stage the planning and design, to the construction and maintenance, we should fully integrate the concept of ecological and environmental protection and advocate ecological transportation infrastructure construction. in the planning and design stage, according to the regional ecological carrying capacity, we should reasonably optimize scale and layout of transportation infrastructure network and avoid "forbidden areas";

In the construction phase, we should promote the application of energy conservation and emission reduction technologies like mixing asphalt under a certain temperature, asphalt cold regeneration, boost the applications of snowmelt asphalt pavement, such as low noise pavement and other functional road technology. We should implement energy conservation and environmental protection at expressway service areas and promote the applications of photovoltaic power generation, LED lighting, energy-saving power supply, sewage disposal and recycle and some other energy conservation and emission reduction technologies. In maintenance phase: we should reasonably design transportation infrastructure maintenance cycle, choose curing technology ecological and environmental friendly. We should eliminate maintenance devices with high energy consumption and high emission and encourage post-assessment on the ecological environment impact of the transportation infrastructures. For transport infrastructure which cannot meet the latest environmental protection requirements because of the early construction idea and technical reasons, we should implement special ecological restoration projects.

Boost Applications of Green Transportation Technologies

Further release potential of traffic field's emission reduction and promote energy conservation and emission reduction with science and technology. At the same time continue to strengthen construction of fundamental ability of statistical monitoring system, measuring system and energy consumption and emission model and other energy conservation and emission reduction.

(1) Further promote technologies and products in energy conservation and emission reduction

Rail transport: carry out energy-saving reconstruction of lighting system transformation and ventilation and air conditioning system, dig out the energy saving, explore into the application of brake absorption technology and select and integrate various energy conservation and emission reduction technologies like environmental quality monitor, energy management system and train renewable energy recycling to form green demonstration points. In the respect of exhaust treatment: promote applications of industry transportation vehicle exhaust treatment, particle trap, fuel additives, etc and strengthen the effective application of efficient emission—reducing devices. Ecological driving: study the driving behavior modification techniques and effect evaluation methods, conduct ecological auxiliary driving system research and development, and select enterprises to carry out the demonstration. In the green maintenance aspects: establish and perfect the green automobile industry management system, promote green motor repair technology and form sets of process specifications with less maintenance waste and harmful emissions and high resource utilization.

(2) Keep enhancing construction of infrastructural capacity

Complete the construction of transport energy conservation and emission reduction statistical monitoring system, build Beijing's transportation energy conservation and emission reduction statistics and monitoring platform (phase II), promote the construction of the energy management platform at enterprise level, achieve interconnection between business platform with government, establish a statistical monitoring system covering energy consumption, fuel and emissions, carbon emissions and other indicators, radiating the Beijing-Tianjin-He- bei area. Push for transport industry measurement standardization, promote advance installation of vehicle-mounted measuring instruments, gradually realize complete coverage of industry vehicle monitoring measurement and make key breakthrough in new energy vehicle and facility efficiency testing.

Gradually improve Beijing's transportation environment monitor network system, enlarge the coverage of online monitoring, strengthening the construction of transport energy conservation and emission reduction laboratory and improve the capacity of traffic environment monitoring. Carry out technology research and development and the system construction of 4-level traffic comprehensive energy consumption and emission model, and make breakthrough in the traffic pollution control integration, the contribution rate of the motor vehicle exhaust pollutant, real- time dynamic energy consumption and emission evaluation on urban road network traffics and traffic pollution index.

Cultivation of Consciousness of Green Behavior

Traffics relate to the daily life of people's livelihood, so the active participation of the whole society is required to promote energy conservation and emission reduction. During the 13th five-year plan period we should give full play to the potential of society and the public and form a harmonious atmosphere of the whole society in participating in and common governance. On the one hand, we should increase social consciousness propaganda of green low-carbon transportation, at the same time, facing vehicle drivers in the city, we should carry out the ecological driving training and promotion and popularize ecological driving concept and technology. In addition, we should strengthen the enterprise's main body responsibility consciousness, through policy promotion, corporate training and other forms, promote enterprises to carry out the energy conservation and emission reduction technology innovation, research and development and application, encourage enterprises to independently develop construction of energy management system, and configuration and other measuring instruments, and actively participate in carbon trading, energy audit and cleaner production audit work. At the same time, we should strengthen the cooperation and communication with the international -related organizations and give full play to the advantages of technical resources and international channels. Eventually we'll make the whole society working together to promote an environment and atmosphere of energy conservation and emission reduction.

Green Transportation Management System

During the 13th five-year plan period, Beijing's transport energy conservation and emission reduction work will enter into the phase of construction of integrated management system and improve comprehensive management ability to work through perfecting the policy system, innovation of systems and mechanisms.

We should strengthen source management of energy conservation and emission reduction, implement the demand management policy of integration of pollution control and jamming prevention, adhere to the total amount of passenger vehicles under control within 6.3 million by 2020 and increase the proportion of new energy car index. We should make I\integrated use of economic levers and necessary administrative means, effectively reduce the intensity of motor vehicle use, guide outgoing transit heavy diesel trucks and other vehicles to bypass the city roads and reduce emission from transit vehicles.

We should perfect the energy conservation and emission reduction policy system, implement application of green freight policy to push for application of energy saving technology, transport efficiency optimization and enhancement of energy conservation and emission reduction management ability, study and establish the green tourism coach enterprise development evaluation index system and incentive policies, establish, improve and implement vehicle inspection and maintenance (I/M) system and standard system and improve the ability of vehicle exhaust control. We should research on monitoring and evaluation policy in operating period as to traffic projects.

We should complete the relevant standard for reduction of energy conservation and emission, push for the measurement data collection, measuring device test and other respects of operative vehicles as well as prepare and release of industrial and local standards in the respects of energy-using equipment, facilities, transportation enterprise management, energy efficiency and carbon emissions and construction traffic infrastructure and speed up the revision of standards.

We should explore into innovation of green traffic mechanism and system, study the construction of energy conservation and emission reduction legal system and find ways to achieve goals by legislation. We should further make clear boundary and assessment indicator system of traffic energy consumption, fuel, pollution emissions and carbon emissions and implement target decomposition and appraisal system for transport energy conservation and emission reduction.

REFLECTIONS

Transportation Energy Consumption, Pollution and Carbon Emissions Are Closely Related to Each Other and Need to Be Governed as a Whole

Energy conservation and emission reduction work in traffic field involves three areas: energy (fuel), pollutants, carbon emissions and three has close correlation with others, but there are certain effect exclusivity and conflicts of interest under different policy orientations. At the same time traffic moving source and energy consumption and emission show regional transmission characteristics. Based on overall consideration of interests of all parties, we need to build up a regional coordinating mechanism among authorities like environmental protection, development and reform and traffics, as well as different provinces and cities and fully consider the relation among energy, pollutant and carbon emission when making relevant policies and measures for energy conservation and carbon emission reduction and air quality improvement.

Structural Optimization Is an Important Approach of Megacities to Break the Bottleneck for Development towards Green Traffics

As a megacity with population of more than twenty million, restricted by factors such as land, resources and environment, Beijing must give full play to the important role of public transportation, increase attraction by improving public transportation service level, facilitate the implementation of demand management policy, effectively reduce the intensity of use of the cars and make travel structure optimized. At the same time, combining with the supply side structure adjustment strategy, we should promote new energy vehicles, improve emissions standards, make green low -carbon development of vehicle structure and launch new policies to encourage the applications of new clean energy vehicles in private sector. Structural adjustment of traffics and transportation is the basis for governance of pollution and traffic in a megacity.

Consolidate Infrastructural Capability to Promote Reduction by Inherent Capacity and Obtain the Actual Effect

Currently the data base of traffic emission energy consumption is relatively weak and the issues like different diameters and unclear base still exist. All of collection of data of mobile source emissions, analysis and forecasting for decision making and massive multi-source heterogeneous data mining analysis need to be solved through construction and perfection of infrastructural capacity.

We need to build a perfect traffic energy consumption and emission statistical monitoring system to master the real level of emissions of mobile source and fixed sources, construct traffic environment monitoring network and control quality conditions surrounding traffic facilities and further explore the effect between traffic and air pollution and influence on human health while using the multi-level comprehensive model to effectively integrate mass multi-source data, strengthening mining, analysis and application, paying attention to effect assessment and quantification of all kinds of energy conservation and emission reduction measures, and supervise and enforce enterprises with high energy saving and high pollution strictly based on objective data. We should reduce the ambiguity of decision-making and promote actual effect from "reduction by inherent capacity"

CONCLUSION

In the background of constructing an international first-class harmonious and livable capital in Beijing during the 13th five-year plan period, a higher view of strategy is necessary to see the conduct of transport energy conservation and emission reduction. All of the pattern selection, strategic policy and goal setting of urban traffic development around need corresponding adjustment around this direction. Energy environment as an external rigid constraint will stimulate the endogenous power for development of transport industry and green environmental protection will lead the transformation and upgrading of Beijing's transport industry in coming five years.

REFERENCES

1. Beijing Statistic Communiqué for National Economy and Social Development in 2015 and during the 12th Five-Year Plan Period.
2. (2015) Annual Report for Energy Conservation and Emission Reduction in Transportation Field in Beijing. Beijing Center for Transport Industry Energy Conservation and Emission Reduction, Beijing.
3. Zong, B.H. (1994) Scenario Analysis Method in Strategic Forecasting. Forecasting, (2), 50-51.
4. Kahn, H. and Wiener, A.J. (1967) The Year 2000. Mac Millan, New York.
5. Beijing Environmental Protection Bureau: Document Released by Beijing Environmental Protection Bureau for Opinions on Local Emission Standard for Vehicles in the 6th Phase in Beijing. http://www.bjepb.gov.cn/bjepb/413526/331443/331937/333896/4380939/index.html
6. Wang, J.W., Mao, Q.Z. and Yang, D.F. (2008) Energy Influences of Urban Planning: An Urban Transportation Energy Consumption Model. China Planning Annual Conference (Dalian).

INDEX

H

I

K

L

M

S

T

U

V

W

Z